Biogeography: A Very Short Introduction

Available soon:

For more information visit our website

www.oup.com/vsi/

Mark V. Lomolino

BIOGEOGRAPHY

A Very Short Introduction

OXFORD
UNIVERSITY PRESS

OXFORD
UNIVERSITY PRESS

Great Clarendon Street, Oxford, OX2 6DP,
United Kingdom

Oxford University Press is a department of the University of Oxford.
It furthers the University's objective of excellence in research, scholarship,
and education by publishing worldwide. Oxford is a registered trade mark of
Oxford University Press in the UK and in certain other countries

© Mark V. Lomolino 2020

The moral rights of the author have been asserted

First edition published in 2020

Impression: 1

Published in the United States of America by Oxford University Press
198 Madison Avenue, New York, NY 10016, United States of America

British Library Cataloguing in Publication Data

Data available

Library of Congress Control Number: 2020932777

ISBN 978-0-19-885006-9

Printed in Great Britain by
Ashford Colour Press Ltd, Gosport, Hampshire

Contents

List of illustrations

List of illustrations

Chapter 1
Biological diversity and the geography of nature

For those trying to make sense of the forces structuring the natural world, biological diversity, the variation in all characteristics of life, from cells to individuals to communities, offers a compelling but often overwhelming source of stimulation and challenge. The challenge lies in the implicit paradox of searching for simple explanations for patterns generated by what are arguably the most complex systems in the universe—the self-organizing biological systems that are the products of billions of generations of engineering during the ecological and evolutionary assembly of our living planet.

Throughout history, key insights into understanding the diversity of life and the processes generating it have come from placing natural phenomena within a geographic context. From Alexander von Humboldt and other visionaries of the Age of Enlightenment, through Charles Darwin, Alfred Russel Wallace, and other naturalist "geologists" of the 19th century, and on to Edward Osborne Wilson, James Hemphil Brown, and other contemporary scientists exploring patterns of variation in life across the planet, all recognized the invaluable clues provided by the geography of life.

For ancient civilizations, their very existence depended on knowledge of how food plants, game animals, and other natural resources were distributed across the landscapes and seascapes that comprised their homelands. Authors across a range of literary genres have spoken of the distinctiveness of place—whether homeland or some exotic island or distant continent. Homer spoke of Odysseus and his travels to exotic islands across the Mediterranean—each with their distinctive, albeit mythical, faunas. Gertrude Stein's *"there is no there there"* is a pointed lament over the loss in the special character of her home town—Oakland, California—after it had lost its characteristic sights, sounds, and smells to join the homogenizing expansion of urban lands.

In the geological sciences, William Smith was the first to unlock the successive build-up of geological formations by discovering the patterns of variation in rock and fossil strata with depth, and horizontally across the landscapes of Great Britain. A global-scale model of the evolution of the Earth's crust would await Alfred Lothar Wegener and his theory of continental drift, which again relied heavily on his abilities to place patterns of variation of rock forms, fossils, and other material in a geographic context. The juxtaposition of geological and paleontological evidence from coasts now on either side of the Atlantic led Wegener at the dawn of the 20th century to one of the greatest discoveries in the history of science. His insight, that the continents have drifted over time, led to the more general theory of plate tectonics—the recognition that the Earth's surface consists of great plates that have been in continual flux since the crust first solidified some 4.5 billion years ago.

The geological evolution of Earth has fundamentally influenced the evolution of its life forms, which brings us back to biological diversity and to the central and unifying assertion of this book and of biogeography in general. Just as the distinguished evolutionary biologist and Eastern Orthodox Christian Theodosius Dobzhansky argued in 1973, that "nothing in biology makes sense except in the light of evolution", biogeography asserts that many if not most of the compelling kaleidoscope of patterns in biological diversity make little sense unless placed in a geographic context. Where do particular species occur? How and why do they vary from place to place? Where did their ancestors occur? How did they spread across the globe? Where are the hotspots of biological diversity? Where can the rarest and most bizarre and most distinct life forms be found today? And how should we apply these insights from the geography of nature to develop successful, spatially explicit strategies for conserving life across the planet? All these questions fall within the realm of biogeography.

As Dennis McCarthy observed in *Here be Dragons*, "Biogeography, once a secret delicacy enjoyed only by geniuses, must now be elevated from its current obscurity and placed alongside literature and history as an indispensable component of a truly enlightened education."

Globalization of the natural sciences

Modern scientists are only recently developing an appreciation for the scientific knowledge of early civilizations, which, at least in terms of the detailed natural history of the local plants, animals, and environments they depended on, may well have exceeded that of many contemporary natural scientists. This traditional knowledge, after all, was essential to their survival. Itself the product of thousands of generations of cultural evolution and natural selection, this knowledge included not only their abilities to identify hundreds of species they relied on, but also the lessons of where to find them—that is, the geography of life, albeit at a local scale.

Early civilizations living in mountainous areas saw, as they turned their gaze toward the slopes, an orderly progression of habitats—each with a predictable yet distinct collection of plants, game species, and other resources. Other populations of our ancestors living in coastal environments developed an intricate and essential knowledge of variation in sea life from shallows to open waters, of the rhythms of the tides and the seasons, and, not only how best to harvest, but eventually how to manage and sustain the resources of land and sea critical for their survival.

Polynesians and other insular civilizations across the world's oceans developed an intricate knowledge of patterns of life across their islands—a knowledge base that continually expanded as burgeoning populations and dwindling resources drove them to explore and expand their realm to distant archipelagos and new regions. Their discoveries were not only essential to the continued existence of their local populations, but simultaneously expanded their knowledge of the geography of life from local to regional levels. During their voyages, Pacific Islanders no doubt observed patterns among islands that form much of the empirical framework for the modern field of island biogeography—that diversity of plants and animals is higher in larger systems, and that more isolated islands tend to have fewer species and many novel forms not found in their homelands. Polynesians who colonized the most remote systems may well have observed a pattern fundamental to the modern field of biogeography—that different archipelagos, even those with similar climatic and environmental conditions, were inhabited by different plants and animals.

This early knowledge of the natural world was foundational to the natural sciences including biogeography, evolution, and ecology. But for this traditional, ancestral knowledge—developed by a mosaic of civilizations—to coalesce into a genuine *science* of the geography of life, we had to advance along two transformative fronts. First, we had to expand the scale of that knowledge and of our views of nature from local to regional and global levels. And second, we had to develop a conceptual framework of fundamental principles and unifying theory to guide and integrate those advances.

Naturgemälde—von Humboldt's holistic portrait of nature

Alexander von Humboldt was one of the world's most remarkable, visionary, and integrative scientists, arguably unrivaled in his holistic view of all human endeavors, including philosophy, art, politics, and government and, of course, the natural sciences during the critical, foundational period of the 18th century—the Age of Enlightenment and the Century of Philosophy. His books and, in particular, the narratives of his travels across the Atlantic and through the tropics of South America were essential reading for all intellectuals across Europe during the late 18th and early 19th centuries, including Charles Darwin, Alfred Russel Wallace, and a legion of other highly impressionable, budding naturalists who were searching for adventures to distant lands in their quest to unlock the mysteries of the natural world.

Among all his numerous and visionary contributions, it may be folly to single out any one of von Humboldt's discoveries as most influential in the history of science. Indeed, his greatest influence on advancing science, in the broadest sense, was not in any particular contribution, but in his *approach* to all natural phenomena—an approach captured in the German term *Naturgemälde*. As von Humboldt applied it throughout his life, *Naturgemälde* directed naturalists and other scientists to view and portray nature in its most holistic form: as a constellation of interconnected phenomena whose patterns can best be envisioned, and whose causes only understood, when explored in synergy. Von Humboldt's *Tableau physique* of elevational gradients along the slopes of Mount Chimborazo in Ecuador (Figure 1) is a prototypic exemplar in biogeography, where somehow he simultaneously demonstrated his dedication to meticulous, painstaking detail with his singular visions for grand synthesis across local to regional phenomena. Because of his pioneering influence during this foundational period in the

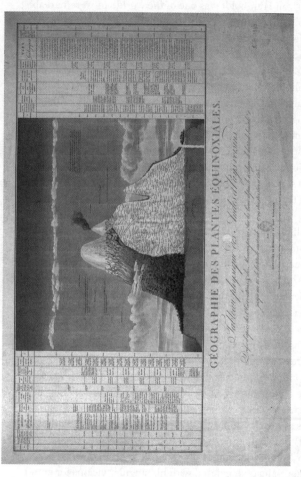

1. Alexander von Humboldt's 1805 *Tableau physique*, illustrating his holistic view of nature and how a variety of interdependent environmental and biotic factors vary along an elevational transect up Mount Chimborazo, Ecuador. Factors listed in the columns to the right and left of the illustration include elevation; distance that mountains are visible from the sea; electrical phenomena; azureness of the sky; humidity; temperature, air pressure, and chemical composition of the air at this elevation; intensity of light and temperature at which water boils at this elevation; scale at which particular plants and animals live; and cultivation of the soil.

natural sciences, von Humboldt was widely recognized by subsequent generations of scientists as the founder of a long and eclectic list of disciplines including volcanology, oceanography, anthropology, archeology, meteorology, geomagnetism, and, most relevant to this book, phytogeography (the geography of plants).

Von Humboldt's *Tableau physique* is also an exemplar in biogeography because it is a compelling illustration of our central assertion—that some of the most complex and potentially confounding aspects of the natural world, and of biological diversity in particular, are rendered explicable when placed in an explicit, geographic context. In this case, the unifying template for von Humboldt's holistic vision of climate, soil, plants, animals, and impacts of humanity was geography—specifically, the elevational gradient in all of these properties and processes, which revealed their interconnectedness and the emergent patterns at the grand scale of the entire slope of Mount Chimborazo.

Epiphanies of geography and evolution

As singular as von Humboldt was in his unique combination of drive and diversity of pursuits, other luminaries in the history of science took an analogous approach to achieve their own epiphanies on the dynamics and complexities of the natural world. Because they studied all the earthly phenomena, natural scientists of the 18th and 19th centuries were often called "geologists." Today we tend to view geology as a science separate from the life sciences; a specialization and compartmentalization of the natural sciences that likely would have vexed von Humboldt. Indeed, some of the most important discoveries in geology were achieved by early naturalists researching a broad array of changes in the structure and composition of rocks and the fossils they contained.

As alluded to earlier, William Smith was another luminary in the history of science, but in stark contrast to the wealthy and aristocratic von Humboldt, Smith's life was a long saga of debt and destitution, punctuated however with another grand synthesis of how nature works—another epiphany derived from placing a vast collection of local observations in an explicit, geographic context. In this case, the observations were on the form and composition of rocks and fossils recorded while Smith surveyed for the mines and canal agencies, and the geographic gradient was not elevation, but depth and distance across the landscapes of Great Britain. Smith was the first to describe and visualize the repeated sequence of rock strata along these two geographic gradients, and to envision how these formations over space chronicled the development of geological formations over time—illustrating all this in a 74 by 105 inch map that would later and justifiably be described as "the map that changed the world" (Figure 2).

Three other exemplars of epiphanies gained by placing complex natural phenomena within an explicit, geographic context are those of Charles Darwin during the early to middle decades of the 19th century, of Alfred Russel Wallace later in that century, and of Edward Osborne (E. O.) Wilson during the 20th century. Darwin and Wallace share credit for simultaneously, but independently, articulating the theory of evolution by natural selection. To Darwin, the essence of the theory came to him, as the story goes, after regrettably mixing specimens of tortoises from different islands in the Galápagos. He was rescued by a resident of the islands who could identify the source island of each specimen by the peculiarities of its carapace. In isolation on the islands, each with its own, distinct environments and selection pressures, those phenotypes best adapted survived and reproduced on that island, ultimately resulting in evolutionary divergence in forms among the islands.

Alfred Russel Wallace initially failed in his first attempt to follow in von Humboldt's footsteps as a naturalist-explorer—collecting specimens of butterflies and birds of tropical South America. After the devastating loss of his specimens to a shipwreck in the Caribbean, the young and indefatigable Wallace turned despair

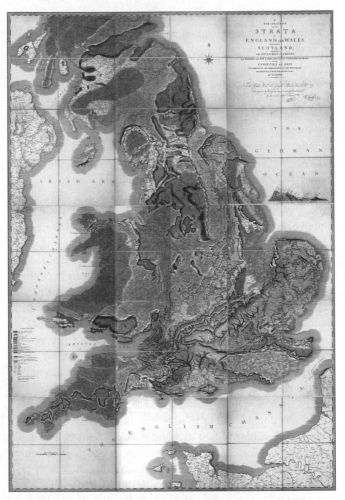

2. William Smith's "map that changed the world" illustrating the relationship between the horizontal and vertical zonation of rock strata, and their temporal, depositional sequence across England. Uplifting and erosion has exposed rock strata along a northwest to southeast gradient of decreasing antiquity (i.e. oldest rock strata lie in the upper left-corner of Smith's seminal map).

into triumph by forging off on his life-defining explorations across the Malay Archipelago (Indonesia). Here he spent some eight years exploring the islands, documenting and describing new forms of plant and animal life, describing their variation among islands, and experiencing epiphanies that led to his independent discovery of the theory of evolution by natural selection, along with a long list of other revelations of how life evolves over space and time. Whereas Darwin spent much of his time and energies on other aspects of evolution, long-distance dispersal, and related subjects, Wallace's central focus was on the geographic variation of nature and the processes driving that variation.

Among Wallace's many contributions to the field that would lead him to be recognized as the father of zoogeography, one visual exemplar first published in 1876 revolutionized biogeography; and we continue to use it today. Following observations by earlier geologists and naturalists that different continents are inhabited by assemblages of different species (biogeography's first principle, discussed below), Wallace realized that each region was a distinct evolutionary arena—each developing its distinctive fauna and flora in isolation and over the evolutionary history of that region and lineage. Like William Smith's exemplar, Wallace's most broadly recognized contribution was another map, but this time one at a global scale and depicting the evolutionary divisions ("zoogeographic regions") of the world (Figure 3). Over the subsequent decades and up to contemporary times, modern biogeographers have sought to refine Wallace's seminal map and to extend it to other fauna besides vertebrates and to plants and other taxa as well, but his vision of the evolutionary divisions of the world continues to shape our views of how life evolves over space and time.

Other exemplary case studies in applying the biogeographer's macroscope for discovering the forces driving patterns in biological diversity are far too numerous to list here, and likely in longer books on biogeography, ecology, and evolution. But one other

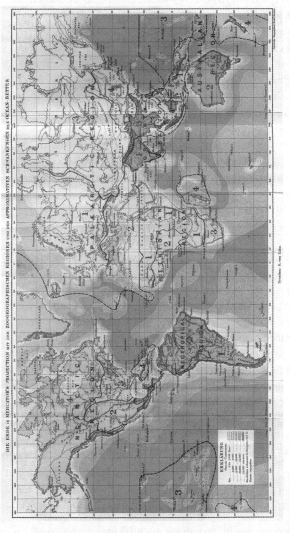

3. Alfred Russel Wallace's 1876 map of the world's zoogeographic regions, showing hierarchical divisions of regions into sub-regions, along with key topographic barriers marking these evolutionary divisions of the world. Principal regions include Nearctic (North America and Greenland), Neotropical (South America), Palearctic (Europe and central to northern Asia), Ethiopian (Sub-Saharan Africa and Madagascar); Oriental (India, Southeast Asia, and Indonesia to the Island of Bali) and Australian (the Island of Lombok eastward to New Guinea, Australia, Tasmania, Melanesia, Micronesia, and New Zealand).

epiphany, in fact one that involves all three disciplines, deserves our attention. It is not hyperbole to describe E. O. Wilson as one of the most influential biologists of the 20th century. He pioneered studies of the systematics, distributions, and ecology of ants; articulated a new synthesis of behavior and evolution—Sociobiology; co-developed one of the most influential theoretical models of biological diversity of islands; popularized the term "biodiversity"; and, as much as anyone else of his time, he became an ardent and inspirational champion of the then nascent field of conservation biology. Indeed, Wilson clearly articulated the connections among biogeography, ecology, and conservation biology when he observed that "The vastness of the tropical archipelagos also provided the knowledge Wallace needed to conceive the biological discipline of biogeography, which has expanded during the late twentieth century into a cornerstone of ecology and conservation."

One of Wilson's most visionary and integrative epiphanies came from a series of maps he compiled to describe the distributions of ant species across the islands of Melanesia (archipelagos east of New Guinea and Australia). In his 1994 book *Naturalist*, Wilson described how "one January morning in 1959 while I sat in my first-floor office ... sorting my newly sketched maps into different possible sequences—early evolution to late evolution. . . . I knew I had a candidate for a new principle of biogeography." Wilson's new principle was his **Taxon Cycle Theory**, which describes a regular series of ecological, evolutionary, and biogeographic changes from colonization of an island by a particular species to its ultimate demise and replacement by new colonists (see Chapter 5).

Fundamental patterns, processes, and unifying principles

The first principles of biogeography are intricately linked to its most fundamental patterns, yet the mutualistic relationship between discovery of patterns in nature and the inference and articulation of fundamental principles from those patterns is seldom appreciated. The observations of the early naturalist-explorers were guided by their preconceptions, and it was through the iterative processes of assessing and refining those preconceptions that the conceptual framework (hypotheses, predictions, theory, and first principles) of biogeography developed. As I like to muse to my students, empirical observations are essential, but without theory and first principles we are destined to walk randomly through the forests bumping into the trees.

Perhaps surprisingly, the first principles of biogeography and articulation of its most fundamental pattern far pre-date those developed by Wilson in the 20th century, and even those proposed by von Humboldt and later by Wallace in the 19th century. The field's first principle came, not from a validation, but from a violation of a then universally held preconception.

Georges Louis le Clerc Comte de Buffon was a contemporary of Carolus Linnaeus in the 18th century. Both viewed science as a means to serve the Creator by exploring and understanding His design of the natural world. The prevailing view of nature was that all life forms were perfectly adapted to their local environments, thus generating the expectation that regions with similar environments should be inhabited by the same species. Buffon was the first of the naturalist-explorers to systematically reject this preconception and describe the distinctiveness of regional biotas. His observations on the distinctiveness of assemblages of mammals and birds of the tropics in South America versus those in Africa would soon be verified by other naturalists-explorers for other groups of species and for other types of ecosystems as well.

Buffon's observations, first articulated in the middle decades of the 18th century, would become biogeography's most fundamental pattern and its first principle—now known as **Buffon's Law**:

different regions of the globe, even those with similar climatic and other environmental conditions, are inhabited by different species. Buffon's Law is fundamental to biogeography because it describes the distinctiveness of place in terms of its life forms. It served as the field's first, unifying principle because it led to the elucidation of the interacting processes that together explain how each region serves as a distinct, evolutionary arena; how life forms that have evolved in one region can disperse and invade another; and how the diversity of life forms is often checked by extinction.

It thus follows that Buffon's Law would eventually lead us to biogeography's fundamental processes of **evolution**, **dispersal** (**immigration**), and **extinction**—and to the unifying principle that *all patterns we study in biological diversity across space and time are the products of these three processes.* Many other factors and processes may have influenced the distributions and diversity of life on this planet, but their ultimate influence on patterns in biological diversity is through their effects on how species and lineages evolve, how they or their descendants disperse to other regions, and how long they survive before ultimately suffering extinction.

One final layer of complexity is necessary to complete this overview of the conceptual framework of biogeography, and it is illustrated in the model shown in Figure 4. Biological systems are among the most complex systems we can study, not only because they are the products of millions of generations of evolution, immigrations, and extinctions occurring across scales ranging from local environments to entire continents and ocean basins, but because each of the fundamental processes, at all temporal and spatial scales, is affected by ecological feedback. That is, the species are not only the products of the fundamental processes, but through competition, predation, parasitism, mutualism, and other more complex multi-species interactions they strongly influence how other species evolve, disperse, and survive or suffer extinction.

14

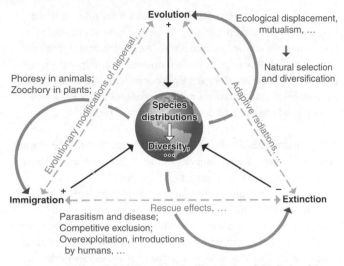

4. A conceptual model of biogeography illustrating the relationships among its fundamental processes (immigration, extinction, and evolution), which either act directly (solid, straight arrows), interact with each other (dashed arrows), or are impacted by interspecific interactions (curved arrows) to shape the distributions of particular species and the geography of life, in general. Phoresy and zoochory describe the processes of dispersal of small animals by larger animals, and of plants by animals, respectively.

Natural experiments and the comparative approach

Biogeographers have tended to be very visual in their approach to exploring nature, with maps serving as the key means of discovering patterns, especially during the early history of the field. As biogeography advanced, we developed more diverse and creative ways of visualizing patterns including employing computer technology in Geographic Information Systems (GIS) and statistical approaches for analyzing and testing predicted patterns in the geographic variation of life. Much of this progress proved the old adage that "necessity is the mother of invention." Despite the compelling nature and insightful powers of our discipline, to be a biogeographer means we are often limited in our experimental approaches. Whereas scientists in other disciplines can control and manipulate experimental treatments in petri dishes, flower pots, microscope slides, or populations of laboratory specimens, biogeographers can seldom if ever conduct experimental manipulations at the spatial and temporal scales most relevant to the broad-scale patterns we study.

As my distinguished colleague and most influential mentor James Hemphil Brown often observes, in biogeography manipulative experiments are either impractical, impossible, or immoral. Seldom can or should we fundamentally alter an entire biota, destroy a mountain, reduce the size of an island, or connect long-isolated islands, continents, or oceanic basins. What we can do, although lacking in control of manipulations, offers far greater realism in that we can study the patterns at scales most relevant to the processes believed to be in play (evolution, immigration, extinction, plate tectonics, etc.). These are processes that operate at broad spatial and temporal scales far beyond those captured in a petri dish, greenhouse, or laboratory experiment. So it is that out of the necessity of seldom being able to employ manipulative experiments, biogeographers have developed myriad creative approaches for testing their hypotheses and advancing theory on how nature works. These include strategies for designing natural experiments, alternative applications of the comparative approach, and an entirely new field and approach to exploring patterns in nature called **macroecology**.

Great lessons of Earth and the road ahead

It should be clear by now that the field of biogeography is one of the most holistic disciplines of biology and, indeed, of all of the natural sciences. While the scope of this highly integrative field brings great challenges in our attempts to combine and synthesize insights across such apparently disparate fields as genetics, geology, paleontology, geography, anthropology, meteorology, oceanography, and ecology, its principles and approaches provide an unrivaled opportunity for understanding biological diversity across all scales of space and time. Biogeography provides an unparalleled opportunity to explore the "great lessons of the Earth."

These great lessons are featured in each of the subsequent chapters. Chapter 2 introduces the concept of the **geographic template**—the very regular (highly non-random) spatial patterns of variation in environmental conditions across the planet from local to global scales. The response of species, each with distinctive physiological tolerances and niches, generates emergent patterns in distributions and diversity of ecological communities across the principal geographic gradients of latitude, elevation, depth, area, and distance (or isolation). Yet despite the regular nature of the geographic template at any particular time, Earth is a dynamic planet that has undergone geological upheavals throughout the 3.5 billion year history of life. Beyond the temporal dynamics in land and sea created by plate tectonics, species assemblages have repeatedly had to adapt to more rapid upheavals; most influential in recent periods were the climatic cycles of the "Ice Ages" (the Pleistocene Epoch, from 2.6 million to 12,000 years before present).

Chapter 3 returns to our central lesson and theme of biogeography—that place matters and that each region, in many cases each island, ocean basin, or lake, can serve as an evolutionary arena, ultimately producing its own biota, distinct from all others. Here we explore the fascinating phenomenon of adaptive radiations and, by applying the comparative approach across four classic case studies, we will see how diversification of these species assemblages is driven by the combined effects of the fundamental biogeographic processes, which in turn co-vary along geographic dimensions (e.g. area and isolation), and are driven by ecological interactions among the species.

Chapter 4 returns to the assertion that evolution occurs across space as well as over time, and presents an overview of approaches used to reconstruct the geographic and evolutionary history of lineages. This overview of the subdiscipline of **historical biogeography,** including the more recently developed approach for mapping lineages across space and time—**phylogeography**—brings us full circle back to Buffon's Law, in this case explored using statistically rigorous approaches for delineating evolutionarily distinctive regions of the globe.

Chapter 5 focuses on biological diversity, its implicit meaning, alternative measures of diversity, and the very general patterns in diversity across the principal geographic dimensions. Here we also demonstrate how research on diversity gradients has its roots in paleobiology—as some of these patterns are quite ancient—and how our exploration of empirical patterns may also provide key insights for those attempting to conserve biological diversity long into the future.

Whereas Chapters 3 through 5 address **macroevolution** (diversification above the species level) Chapter 6 focuses on patterns driven by natural selection and evolutionary diversification among populations (i.e. within a species), or **microevolution**. These patterns include geographic gradients in the anatomical and physiological characteristics of species populations, which in turn are driven by variation in natural selection among regions and across the geographic template.

After presenting an overview of some of the most general ecogeographic gradients across land and sea, we focus on the marvels of island life—genuinely astounding changes in the size, form, and function of species after they colonize islands—changes that all too often lead from the marvels to the perils of island life and wholesale extinctions once oceanic islands are colonized by populations of our own species. Chapter 7 develops our final, "great" lessons: that the global expansions of our own species were strongly influenced by the same factors that shaped the expansions of other life forms; that our indigenous populations were also strongly influenced by the forces of natural selection—driving microevolution of human populations across the geographic template; and, the final lesson, that in our rise to become the world's dominant ecosystem engineer, we have conducted a global-scale manipulative experiment of unsurpassed magnitude. The sobering products of humanity's engineering of Earth's ecosystems appears to be the creation of what some view as a dystopic era of life, the Anthropocene, characterized by waves of species extinctions across land and sea, along with the dissolution of the geography of life and biogeography's most fundamental pattern—the biological distinctiveness of place.

Chapter 2
Dynamic maps of a dynamic planet

Life's geographic template

In Chapter 1, Earth's geographic template was defined as the very regular patterns of variation in environmental conditions across the planet from local to global scales. This is a foundational concept because all biogeographic patterns derive from the responses of species, lineages, and entire biotas to the underlying geographic patterns of environmental conditions that shape natural selection as it varies from place to place. The two component patterns of non-random variation across the geographic template are **spatial autocorrelation** (the tendency for similarity in environmental conditions between sites to decrease with distance between those sites); and **enviro-geographic gradients** (particular trends in environmental conditions along the principal geographic dimensions—latitude, elevation, depth, area, and distance or isolation).

Ultimately, all of the many non-random but diverse patterns of environmental variation that form the geographic template are products of Earth's three great engines. The first is driven by energy from the sun, which not only provides the energy used by nearly all biological communities but also is the principal force driving climatic conditions across land and sea. The second great engine is fueled by the immense heat generated in the Earth's core, which drives plate tectonics and associated geological phenomena. The third engine is powered by gravitational and other astronomical forces that control Earth's revolution on its axis, it's rotation about the sun, and myriad other phenomena including the tides, ocean currents, and spectacular displays such as the aurora borealis and australis, all of which influence activity rhythms for most forms of life on the planet.

Geography of ecological communities

One of the most demonstrable biogeographic effects of the first engine—solar radiation—is the patterned distribution of major types of habitats across the globe. Previously referred to as "life areas" or "life zones," **biomes** are distinguished by their principal forms of vegetation (e.g. tropical rainforests, temperate rainforest, hot desert, and tundra), which in turn form under particular climatic conditions and soils. And here is the connection back to the great solar engine. Consider first how climatic conditions might vary across the Earth if it was flat. In such a planar (albeit impossible) planet, the geographic template would lose most of its character. Latitude would be rendered meaningless because, at least most relevant to our interests here, the intensity of solar radiation would be the same across the planar Earth.

Let us return to the real, spherical world, before the thought experiment gives us vertigo. Because the Earth's surface is curved, solar radiation varies in a highly non-random manner—being most direct and therefore most intense in the tropics, and decreasing in intensity as we move toward the poles. Yes, a simple lesson in geometry and earth science understood by most schoolchildren, but one that can provide some especially profound insights when extrapolated to more fully understand the geography of climates and ecological communities.

The latitudinal variation in solar radiation not only determines concordant changes in annual temperatures from the Equator to

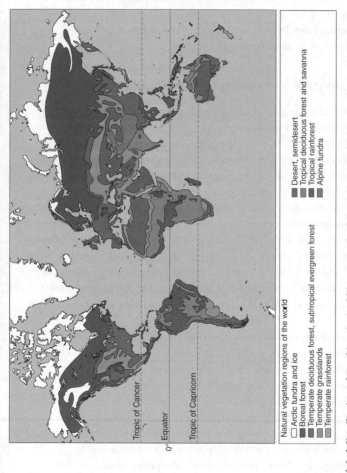

5a. The global distribution of Earth's principal biomes, each of these distinguished by the distinctive growth forms of their vegetation, which in turn is influenced by regional climates and soil conditions.

Natural vegetation regions of the world

☐ Arctic tundra and ice
■ Boreal forest
■ Temperate deciduous forest, subtropical evergreen forest
■ Temperate grasslands
■ Temperate rainforest

■ Desert, semidesert
■ Tropical deciduous forest and savanna
■ Tropical rainforest
■ Alpine tundra

Tropic of Cancer

0° Equator

Tropic of Capricorn

22

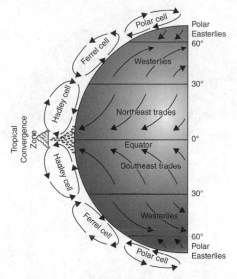

5b. A diagram of atmospheric circulation patterns and their effects on distributions of two principal types of biomes (tropical rainforests and deserts, c and d).

the poles, but it also establishes imbalances in heat energy that drive precipitation patterns, wind currents over land and sea, and, ultimately, the combination of factors determining the regular distribution of biomes across the globe (Figure 5a). First, let us start at the Equator. Here and throughout the tropics, intense solar radiation creates hot air masses that rise above the surface and high into the atmosphere (Figure 5b). As they do so, the pressure (or column of air above them) decreases, and the air masses therefore cool. Because colder air cannot hold as much moisture, water condenses and heavy precipitation occurs across the tropics—this, combined with the intense solar radiation and the lack of substantial seasonality in these conditions, creates Earth's most productive and species-rich biomes—tropical rainforests (Figure 5c).

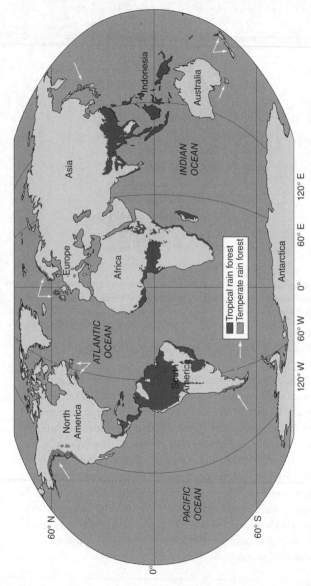

Biogeography

5c. Global distributions of principal rainforests, including tropical rainforests and temperate rainforests (located in mountainous regions along the coasts; marked with white arrows).

We can follow these same air masses, which originally rose up from the tropical rainforests, to understand the forces creating another principal biome—hot desert (Figure 5d). Eventually, those tropical air masses will reach the upper atmosphere and, because they are continually being displaced from air masses rising from the tropical forests below, they split to form two cells of currents ("Hadley cells") flowing north or south. Eventually these air masses cool, become more dense and descend and, as they do, the column of air above them becomes more extensive and, therefore, air pressure increases. That pressure causes temperatures in the descending air mass to rise, which again increases their capacity to hold moisture. Remember, however, that these air masses already lost most of their moisture as they were rising above the tropics. So when they descend, these air masses are not only hot, but also very dry. This is why some of Earth's most expansive deserts are found where air masses from Hadley Cells descend—around 30 degrees north and south latitude.

We can continue this exercise to explain the distributions of most of the world's biomes, but there exists substantial variation in the geography of biomes beyond what can be explained by the solar engine and latitudinal gradients in heat and light. The actual distribution and extent of each biome is also strongly influenced by geological and topographic features. For example, where mountain ranges occur along the coasts and the prevailing winds are from the oceans inland (Figure 6a), air masses rising along the mountain slopes often create rainforests (tropical or temperate) on the windward (ocean-facing) slope (Figure 6b), but once the air passes over the peak and descends it promotes the formation of deserts on the leeward side (Figure 6c) (just as described above for the descending Hadley cells, albeit at a more local scale).

Proximity to the coast, even in the absence of coastal mountain ranges, may also strongly influence local to regional climates. Because of the high heat capacity of water, oceans tend to vary less in temperature compared to land surfaces or the air above them.

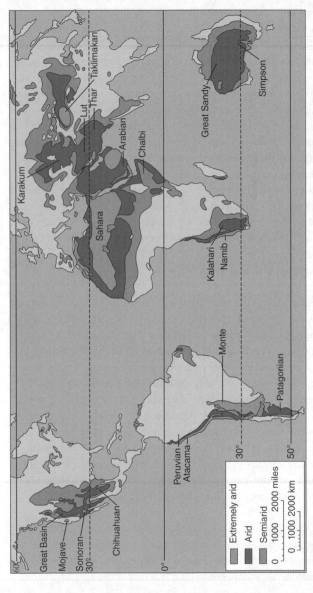

5d. Major deserts (principally located near 30 degrees N and S latitude and in the continental interiors (i.e. distant from oceanic sources of rainfall).

(a)

Elevation (km)

2

1

0

20°   10°   4°   14°   24°

No condensation,   Condensation,   No condensation or evaporation,
rising air cools   rising air cools   descending air warms
10° C/km   6° C/km   10° C/km

6a. Local climates may be strongly impacted by mountain ranges, especially if they occur along the coasts, creating a marked contrast between mesic forests on the windward side, and dry habitats in the rain shadow on the leeward side of the mountain. In this diagram, the ocean is located on the left and prevailing winds blow from left to right.

6b. A montane rainforest.

Thus, coastal environments and those on oceanic islands tend to be thermally buffered, experiencing more mild summers and winters and less seasonal variation in both temperature and precipitation. In contrast, the interior regions of the continents

6c. A rain shadow desert of Puerto Rico.

tend to experience much greater seasonality in climatic conditions and, while typically not as dry as deserts, receive less rainfall than required to support forest biomes. This explains why many continental interior regions such as those of North America, Asia, and Africa are, or at least were once, covered with great expansive prairie, savannah, and other grassland biomes (grasses require less precipitation than forests).

Mountains provide another illustration of the influence of climatic conditions on the geography of biomes, similar to but in this case at a more local scale than the latitudinal distributions of biomes described above. In fact, along mountain slopes in the tropics we often find a sequence of biomes from the foothills to the summits analogous to what we would experience traveling much greater distances from the tropics to the poles (Figure 7a and 7b). As we ascend the mountain, air pressure decreases, temperatures decrease, precipitation increases (at least up to the higher elevations), and local climatic conditions transition along the elevational gradient to mimic those from the tropics to the poles.

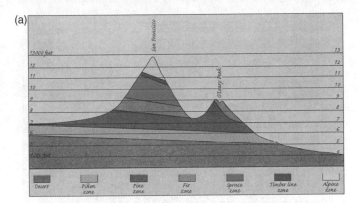

(a)

13000 feet						13
12						12
11						11
10						10
9						9
8						8
7						7
6						6
5						5
4000 feet						4

| Desert | Piñon zone | Pine zone | Fir zone | Spruce zone | Timber line zone | Alpine zone |

7a. **Because pressure, temperatures, and precipitation change as we ascend a mountain slope, the principal types of vegetation also change, often following a regular sequence similar to that exhibited by biomes along the latitudinal gradient from the tropics to the poles (illustrated here in C. Hart Merriam's classic depiction of zonation of "life zones," roughly equivalent to biomes (a) along the slopes of mountains in Arizona, USA, and (b) across North America).**

Geographic template of the marine realm

So far we have ignored what constitutes most of the biosphere: the marine realm, which covers roughly 70% of the Earth's surface. Even a brief overview of environmental variation across the marine realm is far beyond the scope of this book. However, there are strong parallels between the drivers of distributions of ecosystems in the marine and terrestrial realms.

Except for some extremely rare and singular ecosystems clinging to cold seeps and geothermal vents on the ocean floor, the principal source of energy for both marine and terrestrial ecosystems is solar radiation which, as described above, also creates differences in heat across the globe that drive currents, in this case, across the ocean's surface and to its great depths as well. Whereas the particular types of biomes characterizing a terrestrial region are largely the products of climatic and edaphic (soil) conditions,

(b)

7b. C. Hart Merriam's classic depiction of zonation of "life zones," roughly equivalent to biomes across North America.

oceanic ecosystems are strongly influenced by water temperature and water chemistry. The analogy is, of course imperfect; in particular because of the density and thermal properties of water versus air, and how rapidly light attenuates and pressure increases with depth in the marine realm. Still, we can draw other parallels between the terrestrial and marine systems; in particular, that these environments (climate, soil, air temperature, and atmospheric currents above; water chemistry, temperature, pressure, and oceanic currents below) are modified

30

by geological formations—topographic and bathymetric features of the continents and oceans, respectively. Mountain chains, valleys, great canyons, rift valleys, and trenches, whether rising above or lying deep beneath the surface of the ocean, all are wrought and continually transformed by Earth's internal engine—the geothermal forces that have shaped and reconfigured the planet throughout its 4.5 billion year history.

A kaleidoscopic planet in eternal flux

The geological and topographic modifiers of terrestrial and marine environments were understood to be features of such immense scale that, throughout the history of science, from Linnaeus, Buffon, and von Humboldt, through Darwin and Wallace, and up until the middle decades of the 20th century, none questioned that Earth's geological profile was fixed. Charles Lyell, who was one of Darwin's mentors and ultimately recognized as the father of geology, accepted that land surfaces rose and fell relative to the levels of the oceans (through mountain building, erosion, and melting and freezing of glaciers). But he would certainly have viewed any suggestion that entire continents or ocean basins could drift laterally across the globe as heresy or lunacy, or both.

It wasn't until well after Darwin and Wallace provided their foundational insights on natural selection and the distributions and dynamics of life, and after the scientific revolutions in genetics, evolutionary biology, biogeography, and ecology of the early decades of the 20th century, that geologists and other natural scientists finally realized that the Earth was, indeed, a kaleidoscopic planet, in eternal flux. In retrospect, that the natural sciences—in particular those most dependent on a sound understanding of Earth's geological and geographic template (biogeography, evolutionary biology, and ecology)—could somehow achieve revolutionary advances while holding to the doctrine of the fixity of the continents is indeed remarkable, and speaks to the great minds of that more recent "age of enlightenment" of the early 20th century.

The history of what we today call "continental drift" and "plate tectonics" represents a saga of delayed development on a decadal scale. Alfred Lothar Wegener developed the first comprehensive theory of how the Earth's continental landmasses drifted across the globe, including reconstructions of past configurations of the continents and ocean basins, explanations for the driving forces, and how continental drift was related to the creation of geological features including mountain ranges, great rift valleys, the shapes of coasts across adjacent regions of the Atlantic, and the similarity in rock formations along those now separated continents (Figure 8). First proposed in a paper and then a series of books published and repeatedly edited and updated from 1915 to 1930, Wegener's **theory of continental drift** was either ignored or ridiculed by the scientific establishment throughout his life and

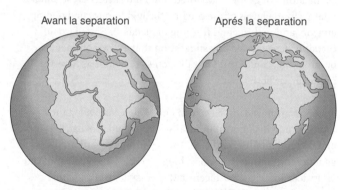

Avant la separation   Aprés la separation

8. **Earth's tectonic plates have drifted across the globe, at times coalescing to form one world-continent (Pangaea) and then splitting and drifting apart, strongly influencing global and regional climates along with the distributions and evolutionary divergence of life forms (illustrations from Antonio Snider-Pelligrini's 1858 reconstruction, which he based largely on the fit of the continents across the Atlantic in contemporary maps—i.e. "Aprés la separation," on the right).**

well after his death in 1930. Ironically, Wegener died during an expedition to further his theory of continental drift by exploring a volcanically and tectonically active region of Greenland. His research team was caught in a terrible blizzard. Wegener made it back to camp but, after learning one of his field assistants was still missing, he set out on his own to rescue him. Neither man was ever seen again.

Wegener's tragic story speaks both to the loss of a singularly visionary scientist, but also to how the advancement of science can be stalled, in this case for some fifty years. His theory of continental drift truly was revolutionary, threatening to overturn the long-held doctrine of the fixity of the continents—a doctrine that was dictated by centuries of scientific as well as religious teachings. Despite the impressive lines of geological and biological evidence amassed by Wegener, this revolution, arguably one of the most transformative in all of science, would require a wealth of information on geologic formations far beyond the reach of geologists of the early 20th century.

The revelations that finally produced the revolution were not the result of direct scientific investigation into the problem but a by-product of the battle among countries to defend and dominate the seas during World War II. The war effort greatly expanded bathymetric mapping across the worlds' oceans, finally (albeit unintentionally) providing marine geologists with the missing pieces that would not only verify Wegener's theory of continental drift, but extend it to a more comprehensive theory on the geological dynamics of our planet. While Wegener's theory addressed drift of the continents, **plate tectonics theory** encompasses all three processes—drift along with the formation and the destruction of crustal plates, simultaneously revealing the mechanisms that drive all three of these interconnected processes.

Perhaps most illuminating of all the various structures revealed by marine geologists were the mid-oceanic ridges; fissures along the

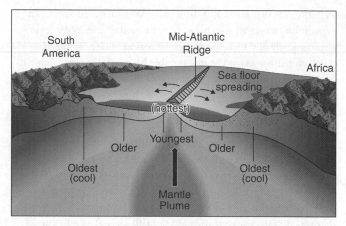

9a. Herman Hess's model of sea floor spreading explains how magma flowing from deep in the mantle rises up to split the oceans apart.

ocean floor that run along its entire length and bisect the oceans. Subsequent geological surveys would reveal that the ridges were the sites where rising magma initially began to rift apart an ancient, oceanic plate, eventually splitting and displacing it with denser, oceanic crust that continued to expand in a zipper-like process called **sea-floor spreading** (Figure 9a). Consistent with the model of sea-floor spreading developed by Herman Hess and his colleagues in 1963, oceanic crust at the ridge is young and relatively hot, with its age increasing and its temperature decreasing in either direction as we move from the magma-driven, mid-oceanic ridges toward the continents (Figure 9b). Where the expanding oceanic crust collides with the less-dense continental crust, it causes upwarping and orogeny (mountain building) of continental crust (such as that which created the Andes of South America) and subduction of the denser, oceanic crust, forming great trenches (e.g. the Marianas Trench, which lies some 11 km beneath the surface of the ocean). Also consistent with the model of sea-floor spreading and plate tectonic theory, continental crust is ancient (much of it dating back to the initial cooling and

Age (millions of years before present)

220 200 180 160 140 120 100 80 60 40 20 0

9b. Herman Hess's model of sea floor spreading explains the age of oceanic crust increasing and its temperature decreasing as we move away from this tectonic hotspot in the middle of the expanding ocean.

solidification of the Earth's surface some 4 billion years ago), while the denser, oceanic crust is continually recycled in great gyres of magma and, thus, is much younger (typically less than 180 million years old).

Although Wegener speculated on the potential role of other forces, he correctly asserted that Earth's geothermal engine creates the energy that drives its turbines—gyres of magma that rise thousands of meters from its core through the mantle and up to the Earth's outer layer of crust. These great gyres of molten rock, in turn, cause the continents to drift, split apart, or collide, while creating new oceanic crust (along continental rift valleys and the mid-oceanic ridges) or destroying more ancient oceanic crust as it is subducted along marine trenches and joins the gyres of magma below. Thus, as with theories in general, plate tectonics theory was formulated on a set of now generally accepted principles—of geothermal energy in Earth's core; great gyres of molten rock rising and circulating through its mantle; different origins and densities of oceanic and continental crust; and the mechanism of sea floor spreading.

In addition to the geological, topographic, and bathymetric phenomena discussed above, plate tectonics also explains earthquakes, volcanoes, the formation of island chains, and a host of related phenomena describing the dynamics of the geographic template—knowledge that, as we shall see in Chapters 3–5 in particular, proves indispensable for understanding the evolutionary development of species assemblages over space and time.

Geological foundations of adaptive radiations

Although islands cover just a small fraction of the Earth's surface, they have played a highly disproportionate role in providing transformative insights in evolutionary biology and biogeography. Arguably the foremost among these arose from Darwin's explorations across the Galápagos Islands, and Wallace's across the Malay Archipelago of Indonesia. Accordingly, our case studies of adaptive radiations in Chapters 3 and 4 will focus on insular lineages including those of the Galápagos and Hawaiian Archipelagos, Madagascar, and the "archipelago" of island-like lakes of Africa's Rift Valley. The geological origins of each of these systems are diverse and often complex, but for our purposes here we need only describe the salient features of their geological development in order to understand the evolution of their inhabitants.

Wegener was correct in postulating that nearly all of the continents were joined together at some ancient time in the past—forming the "world continent" of **Pangaea.** During this period, relatively warm climates persisted across most of the globe, allowing heat tolerant plants and animals to disperse to most regions of Pangaea. The break-up of Pangaea, which began early in the Mesozoic Period, around 240 to 220 million years ago, accelerated the diversification of regional biotas which had now become isolated from the homogenizing effects of gene flow among populations across the supercontinent. Among the major surges in species diversification that occurred in the late-Paleozoic Era was that of the ancient reptiles, soon to be replaced by the dominant vertebrates of the Mesozoic: the dinosaurs, and their more diminutive relatives—mammals, birds, and the ancestors of today's extant reptiles as well.

The southern continents remained together as the ancient subcontinent of "Gondwana," or in close proximity, much longer than those to the north. The long, shared history of Gondwana's landmasses accounts for the relatively high similarity of plants and animals among now distant continents across the southern oceans. Madagascar was set deep in the southern latitudes until it began to drift northward, at first connected and later in close proximity with the then island continent of India (again, their shared geological history explaining some of the remarkable similarity of the Indian and Malagasy fauna and flora, despite their current isolation). By about 80 million years ago, Madagascar reached a position relative to Africa that it retains today, while the Indian subcontinent accelerated in its surge northward across the Equator to collide into Asia some 40 to 50 million years ago—initiating uplift of what today is the world's tallest continental mountain range—the Himalayas.

As alluded to above, mountains also form underwater when magma rises through the oceanic crust, occasionally creating submerged volcanoes (**seamounts**) or oceanic islands if they ultimately emerge above water. The Hawaiian and Galápagos Islands were both created in this manner, as hotspots above mantle plumes that pushed through relatively thin areas of oceanic crust. The temporal dynamics and geological origins of these two archipelagos, however, differ in key respects that have great bearing on the origins and subsequent development of their endemic (unique to these islands) lineages of plants and animals.

The Hawaiian Islands actually are the most recent, distal end of a 6,000 km long chain of oceanic volcanoes whose origin traces back to an underwater region near the Aleutian Islands some 80 million years ago. At that time, an oceanic plate that now forms part of the Pacific Ocean began to expand and drift northward, creating a series of volcanic seamounts and islands extending southward each time a relatively thin portion of the plate drifted over an intense hotspot in the magma below. The ancient geological relatives of the Hawaiian Islands can thus be traced back to seamounts and the eroded, flat-topped **guyots** that were, or are in the process of being, subducted into the Aleutian Trench. From there, the volcanic chain can be traced from the Emperor Seamounts (roughly 80 million years old) extending southward to the Midway Islands, then toward the southeast (following a major shift in the direction of the drifting plate around 47 million years ago), eventually on to the oldest of the Hawaiian Islands above water (Kaua'i, 5.1 million years) and on to the Big Island—Hawaii, which presently lies over the hotspot and thus remains

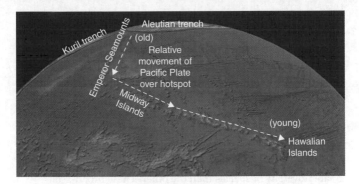

10. The island chains stretching from the Aleutian subduction zone through the Emperor Seamounts and ultimately culminating in the Hawaiian Islands Chain are the results of an 80 million year, conveyor belt-like formation of volcanoes as the Pacific Plate passed over a tectonic hotspot in the mantle below.

volcanically active (Figure 10). Without this complex, but now well-documented geological history, it would be difficult if not impossible to explain some peculiarities of the Hawaiian biota, including genetic and evolutionary research that suggests that some of its lineages are older than the oldest emergent islands (a mystery we return to solve in Chapter 3).

The geological origin of the Galápagos Islands is not nearly as ancient as that of the Emperor Seamount–Hawaiian Islands Chain. The Galápagos, however, were also formed by a hotspot—in this case, one located beneath the junction of three plates—the Nazca, Cocos, and Pacific Plates—currently located near the Equator and some 950 km west of the coast of Ecuador, South America. Although the hotspot may have been active for some 20 million years, the oldest extant island of the principal Galápagos Islands, San Cristobal, is much younger (3.2 million years). From San Cristobal, the ages of the remaining islands decreases towards the northwest, in line with the direction of the hotspot's relative movement, to the youngest principal island, Fernandina—a young upstart at "just" 0.05 million years. Here again, because some antecedents of the current islands are now submerged, we may have temporal anomalies of some Galápagos lineages being older than the current islands. (Seamounts of formerly emergent ancestors of the Galápagos are found to the east and may date back some 10 million years, perhaps substantially longer.)

Chapter 3 will capitalize on these differences in the geological origins, age, and contemporary isolation, size, and topography of these two archipelagos to design natural experiments, applying the comparative approach to explain why one harbors a much greater diversity of species than the other. These broad-scale natural experiments on adaptive radiations across two archipelagos, along with the spectacular radiations of the Malagasy lineages, help elucidate the key geological and geographic drivers of evolutionary diversification. Our explorations of the geography of diversification will, however, include one additional and distinctive system of remarkable radiations—the island-like freshwater lakes of Africa's Rift Valley.

Just as extensive lines of mantle plumes can rip open oceanic crust along the mid-oceanic ridges, the same process can split continental crust to create expansive rift valleys. In fact, they are two stages in the same process. The southern continents of Africa and South America were joined as one continent (a remnant of Gondwana) until around 160 million years ago when a mantle plume caused a rift valley to form. In the subsequent stages, the valley continued to extend as dense, oceanic crust was added to either side of the nascent sea floor, eventually forming the Atlantic Ocean.

The rifting of eastern Africa began roughly 30 million years ago, forming great depressions that eventually filled with precipitation and runoff to create some of the largest and most species-rich lakes in the world. The teaming assemblages of fish and other aquatic organisms in Africa's Great Lakes, however, are much younger than this, providing a powerful illustration of how adaptive radiations can be just as strongly influenced by recent and comparatively rapid dynamics in climatic conditions as by plate tectonics and, ultimately, how the two are interrelated.

The geological drivers of climate

Although the most direct and obvious impacts of plate tectonics on the geographic template and the world's biotas have been its transformations in the size, shape, topography/bathymetry, and isolation (or connections) of the Earth's plates, these dynamics of land and sea have also profoundly impacted climatic conditions across the globe throughout the history of life. To more fully appreciate this interdependence of geology and climate, it is first important to bear in mind that land, whether bare or covered with vegetation, absorbs more solar radiation than the reflective surface of the world's oceans. As a result, during those periods of geological history when the continents happened to be concentrated in the tropics, the Earth's total heat budget was accentuated, triggering a period of intense global warming. Other periods of global warming may have been associated with massive inputs of carbon dioxide, methane, and other natural greenhouse gases released during periods of intensified volcanic activity and tectonic disturbance of oceanic sediments.

Both of these factors (the position of the continents and intensified tectonic activities) may have combined to create one of the warmest periods on record, and possibly the hottest period of the **Phanerozoic Eon**—the 541 million year period bracketing the entire evolutionary history of macroscopic life on the planet. The Eocene Optimum, or mid-Eocene Sauna (~55 million years ago), marks a period when average global temperatures were well over 10° C above those of today, when the Earth was devoid of any significant ice sheets, and when tropical climates and vegetation extended deep into the Arctic and Antarctic latitudes in both hemispheres (Figure 11). The mid-Eocene also marks a period of global-scale turnover of major lineages and taxa, including

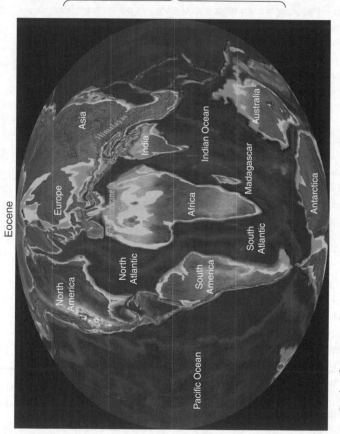

11. During the extremely warm periods of the so-called mid-Eocene Sauna (~50–60 million years ago), tropical and subtropical habitats extended far into the high northern and southern latitudes.

replacement of ancient mammals with the modern orders of today, and a surge in evolutionary radiations of other lineages of animals as well as plants.

The size, relative positions, and configurations of the continents and ocean basins also profoundly impact ocean currents which, in turn, drive or otherwise alter atmospheric currents, thus modifying climatic conditions at regional to global scales. The effect of plate tectonics on oceanic currents is evident from inspection of tectonic reconstructions from the Late Cretaceous Period, approximately 80 million years ago (Figure 12). The gap between the northern continents of Laurasia and the southern continents of Gondwana, which first opened up around 150 million years ago, enabled a circum-equatorial current (the Tethyan Seaway) that drove secondary currents flowing north and south, thus distributing the heat absorbed by tropical waters to the higher latitudes. This accounts for extension of warm climates and tropical vegetation into the higher latitudes, and for the attenuated latitudinal gradient in temperature and the aseasonality of climates that persisted from this period into the mid-Eocene.

Awareness of one further linkage between plate tectonics and climate is important if we are to understand the upheavals in climatic conditions of the Pleistocene Epoch (2.6 million to 12,000 years ago). The twenty or so climatic upheavals of the Pleistocene alternated so rapidly, with periods on the order of 10,000 to 100,000 thousand years, that they cannot be directly attributed to plate tectonics, which operates on scales of many millions of years for substantial drift of the continents and expansions or subductions of oceanic basins. Nonetheless, it appears that plate tectonics did play a fundamental role in the dramatic glacial–interglacial cycles that marked this geologically abbreviated period.

The current distribution of the continents presents a very asymmetrical profile. The area of landmasses in the northern

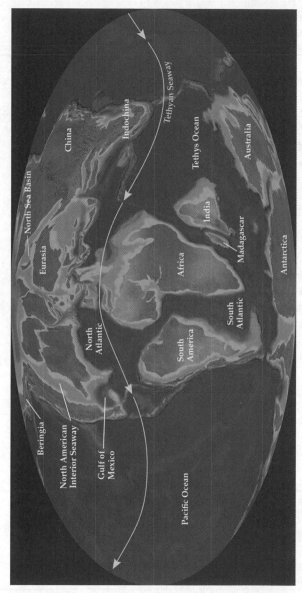

12. Following the breakup of Pangaea, Earth's continental plates drifted far apart and, during the Cretaceous, opened up a circum-equatorial channel (the Tethyan Seaway), which created strong currents that distributed warm tropical waters into the higher latitudes, reducing the latitudinal temperature gradient and seasonal variation in temperatures across most of the globe.

hemisphere far exceeds that to the south. One salient outcome of this complex redistribution of land and sea, and the differential heat budgets north and south of the Equator, is that the Pleistocene became a period of intense if not unrivaled climatic instability. Relatively slight alterations in the solar radiation intercepted by the Earth, which during earlier periods of tectonic configurations were only subtle modifiers of climate, caused the dramatic and repeated cycles of global climate change throughout the Pleistocene.

Climatic upheavals of the Pleistocene Epoch

At the opening of this chapter, I listed the three great engines that drive the Earth's geological and environmental dynamics: solar energy, geothermal energy, and gravitation and other astronomical forces. It is this third engine, acting as a modifier of the first, which created the cyclical changes in solar radiation striking Earth and, in turn, drove the climatic upheavals of the Pleistocene.

Although the total energy emitted by the sun, known as the solar constant, actually varies somewhat over time, the magnitude of this variation is much less than that required to drive the climatic upheavals of the Pleistocene. As I mentioned earlier, these cycles are instead created by cyclical variation in the characteristics of Earth's orbit on its axis and about the sun, altering Earth's total heat budget and the distribution of heat over time and across the planet's highly asymmetrical continental profile north and south of the Equator. The cyclical changes in Earth's orbit include three main features: its obliquity, or angle of rotation, varying from 22.1 to 24.5 degrees every 41,000 years; the orientation of its orbit relative to nearby solar systems, alternating between pointing toward Polaris (the "North Star") currently, and the star Vega every 22,000 years; and third, the stretch or ellipticity of its orbit around the sun, varying from more circular to more elliptical in cyclical periods of 96,000 years. It was the Serbian mathematician and geophysicist Milutin Milankovitch who performed the impressive feat of pulling this all together to provide the first, comprehensive explanation of how these orbital cycles, now known as Milankovitch Cycles, explain the glacial-to-interglacial cycles of the Pleistocene, one of the most climatically unstable periods in the geological record.

The current interglacial represents a relatively atypical period in terms of the prevailing climatic conditions of the Pleistocene. Roughly 90% of the past 2.6 million years was dominated by glacial conditions which, although varying substantially in their magnitude and duration, lasted roughly 100,000 years on average, alternating with interglacials that averaged roughly 10,000 years. The glaciers, which in some regions reached some 2 to 3 km in height, extended deep into what are now temperate latitudes. Far beyond the glaciers, however, climates cooled across land and sea well into the tropics. Air temperatures above the continents cooled by some 4 to 8 degrees C whereas, because of the high heat capacity of water, ocean surface temperatures experienced less cooling ("just" 2 to 3 degrees C, depending on the region and prevailing oceanic currents). Still, these more modest changes in sea surface temperatures had profound effects on corals, aquatic plants, and most other forms of ocean life, and likely impacted regional climates on the continents far inland. The dramatic shifts in temperature and precipitation regimes associated with recent *El Niño* events are often triggered by much more modest shifts in sea surface temperature (roughly at an order of magnitude less than those characterizing fluxes in sea surface temperatures occurring during the twenty to twenty-five glacial periods of the Pleistocene). In summary, dynamics in Earth's orbit during this period profoundly impacted the climatic conditions and related atmospheric and oceanic conditions across the globe.

The climatic reversals were not only profound, but they were extremely rapid—paradoxically, much more rapid than the key drivers of these changes—that is, the cyclical changes in Earth's orbit about the sun (with periodicities ranging from 22,000 to

96,000 years). The Pleistocene's climatological record—stored in ice cores, marine and lake sediments, tree rings, and other chrono-paleoclimatic indicators—reveals that shifts between glacial and interglacial conditions often occurred on the scale of millennia, and sometimes as rapidly as just centuries. These remarkably rapid climatic swings were caused by global-scale, positive feedback phenomena that accelerated Earth's spectral and thermal properties. As the Earth's climate began to cool due to gradual changes in its orbit, ice sheets expanded, increasing the amount of solar radiation reflected back to the atmosphere, thus accelerating the rate of cooling. Alternatively, during a period of global warming, the tables were turned. As glaciers melted, more land was exposed, increasing the amount of solar radiation absorbed, which in turn increased the rate of ice melt. The resultant, self-intensifying accumulation of this greenhouse gas in the atmosphere profoundly accelerated the global warming that punctuated transitions from full-glacier to full-interglacial conditions.

Impacts of climate change on the Pleistocene Biota

Not surprisingly, climate shifts of the Pleistocene deeply impacted biotas across Earth, but the effects varied considerably among taxa. Most, at least those that survived, exhibited large shifts in their geographic ranges—on average, shifting roughly 10 degrees in latitude and, in mountainous regions, species' ranges shifted roughly 1,000 meters in elevation. Given the marked variation in physiological tolerances and niches among species, even within closely related taxonomic groups, geographic shifts in response to climate change during the Pleistocene were often idiosyncratic, in turn causing pervasive reshuffling of biological assemblages. Many species associations that developed throughout the long, more stable periods that preceded this interval were disrupted, often being replaced by novel combinations and assemblages of species not shared with any previous period.

In addition to the sometimes dramatic but frequently unpredictable shifts in their geographic ranges, the Pleistocene biota also exhibited two other qualitatively distinct responses to climatic reversals. Many species that were unable to shift with their drifting optimal climates experienced reductions in their geographic ranges, ultimately resulting in their extinction. Extinctions were especially prevalent among many plant lineages during the initial climatic cycles of the Pleistocene. Those plants, however, that survived this initial filter of climate-driven natural selection tended to persist throughout the remaining twenty or so cycles.

Perhaps most surprising among the possible responses of Pleistocene biota are not only those that survived the initial bottlenecks of climate change, but the many lineages (including those of large terrestrial mammals and birds) that surged in diversity and ecological dominance even during some of the most intense periods of glaciation. The final and, indeed, the most confounding surprise of the ecological dynamics of the Pleistocene is that these same lineages, the large herbivorous and carnivorous mammals, and the large predatory and scavenging birds that fed upon them, collapsed during the final stages of the Pleistocene.

Why should species that grew to such gargantuan size—the **Pleistocene megafauna**—suffer extinctions after surviving the previous twenty or so glacial and interglacial periods? While there is far from universal consensus, the explanation now accepted by a growing number of ecologists and biogeographers is that it wasn't climate change per se that caused the nearly wholesale extinction of the megafauna on most landmasses.

As we noted, while many species suffered range contraction and extinctions, others shifted and expanded their geographic ranges in response to climate change. In fact, range shifts and geographic expansions were often promoted by the onset of glacial conditions which, by locking up such huge volumes of water in the massive ice sheets, lowered sea levels by 100 m or more. This, in turn, transformed shallow seas into land bridges, providing terrestrial life forms with access to otherwise isolated islands and adjacent continents typically separated by shallow seas. Most prominent among these glacial-age land bridges were those connecting the Greater Sunda Islands of Java, Sumatra, and Borneo with the Malay Peninsula and Asia, and that between present-day Siberia and Alaska. The latter, glacial age land bridge comprised the huge, glacier-free subcontinent of **Beringia**, home to a great diversity of megafaunal wildlife, perhaps rivaling that of today's African savannah.

One of the relatively large mammals that utilized both of these land bridges for their immigrations and eventual geographic range expansion across most of the globe was the apparent "culprit" in the global extinctions of the Pleistocene megafauna— our own species. The Malay-to-Sunda land bridge provided our ancestors with access to the large islands of Indonesia, which served as staging posts for our colonization of the Lesser Sunda Islands to the east, and eventually Australia. The first, aboriginal populations of our species reached the glacial subcontinent of Sahul (New Guinea, Australia, and Tasmania) by around 56,000 years ago, and wholesale extinctions of Australia's megafaunal marsupials, birds, and reptiles began within a few thousand years after that (Figure 13). Our ancestral populations reached the glacial refuge of Beringia by around 30,000 years ago, but remained blocked in their expansion into North America until global warming opened a narrow channel between retreating eastern and western ice sheets around 15,000 years ago. Human populations reached regions south of the glaciers by 13,000 years ago, and the North American megafauna began to disappear soon after that, with those of South America and the Caribbean falling like dominoes following successive waves of human colonization (occurring around 11,000 and 6,000 years ago, respectively).

Biogeography

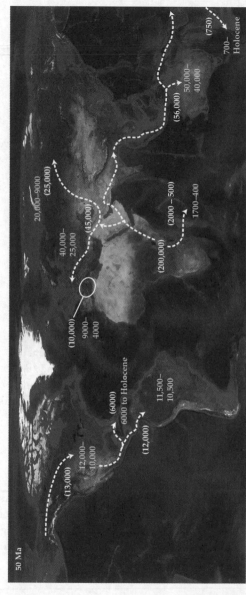

13. Extinctions of the megafauna—huge mammals, birds, and reptiles—occurred during the later stages of the Pleistocene Epoch (in particular, between 50,000 and 11,000 years ago) and followed on the heels of the arrival of ecologically significant humans in those continents (ages of arrival of humans and of megafaunal extinctions in parentheses and in normal type, respectively).

Evidence continues to accumulate in support of megafaunal extinctions being driven by human activities and not climate. Our ancestral populations were remarkable in their abilities to marshal their forces and their collective intelligence to track, hunt, and subdue prey with ingenious weapons and group strategies that sometimes capitalized on local land formations (box canyons, and cliffs) to subdue entire herds of their prey. Eventually, their ecological prowess included strategies for diverting creeks and damming streams, and using fire to transform native habitats into those more optimal for their own populations.

If we consider the alternative hypothesis of climate-driven extinctions, its predictions are just not borne out by the accumulated evidence.

If megafaunal extinctions were caused by climate change, then we would predict that they

1. would have occurred during the first cycles (because they caused climate change so dramatically different from that of the previous periods of relatively stable and more equable climates);
2. would occur alongside those of plant species they depended on for food and habitat;
3. would occur simultaneously on different continents and islands across the globe; and
4. their intensity would be similar across the continents.

Instead, the prevailing evidence reveals that, whereas plant extinctions occurred during the early cycles of climate change, those of the megafaunal mammals, birds, and reptiles occurred during the later cycles (contradicting Predictions 1 and 2). In fact, as we observed above, diversity of megafaunal mammals surged during the early and middle cycles of the Pleistocene. Also, as described above and illustrated in the map of Figure 13, megafaunal extinctions were far from simultaneous, but occurred on the heels of the arrival of humans in each region (in contradiction to

51

Prediction 3). In fact, megafaunal extinctions occurred during glacial maxima on some continents, but during interglacial periods on others.

Finally, megafaunal extinctions varied substantially among the continents. While the megafauna disappeared from most landmasses, those of Africa were largely spared. The most plausible explanation, deduced by my esteemed colleague Jared M. Diamond, is that the African megafauna viewed our species as a familiar ecological associate. Wildlife co-evolved with us in our African homeland, being exposed to and thus able to adapt to each incremental advance in our powers as competitors and predators. For the megafauna of other lands, we were an entirely novel species. Once met with what Charles Darwin referred to as the "stranger's craft of power," the ecologically naive, native biota of lands far beyond our African homeland collapsed, with effects that likely cascaded across the web of interacting and interdependent plants and animals.

Chapter 3
The geography of diversification

Adaptive radiations, the comparative approach, and natural experiments

Adaptive radiations provide compelling and instructive case studies on the geographic factors influencing diversification. Here we will use the comparative approach to design natural experiments to explore how geography influences ecological and evolutionary diversification in the insular systems described in Chapter 2 (the Galápagos and Hawaiian Archipelagos, Madagascar, and the Great Lakes of Africa's Rift Valley). These systems were chosen not only because they include some of the world's most spectacular demonstrations of adaptive radiations, but because they represent compelling illustrations of the **geography of diversification**.

Despite their limitations (in particular the lack of rigorous controls and replicates), natural experiments provide the realism essential for investigating the influence of processes that operate across spatial and temporal scales well beyond that possible for controlled, manipulative experiments (i.e. plate tectonics, evolution, immigration, and extinction). For such experiments, we need to select examples that minimize variation in factors extrinsic to the working hypothesis, and maximize variation in the factors of interest—here, isolation, area, topography, and island age.

Accordingly, let us first compare the adaptive radiations of two similar lineages of birds (both finches) isolated on tectonically and geographically quite disparate systems—the Galápagos and the Hawaiian Islands. We will then switch control and treatment effects by comparing adaptive radiations for two very disparate lineages (honeycreepers and lobeliad plants) within the same archipelago (the Hawaiian Islands). We will adopt a similar approach by comparing the extent of adaptive radiations of different lineages within one of the most intense hotspots of diversity and endemicity in the world—the island-continent of Madagascar.

For our final case study, we will look at the cichlid fish of Africa's Rift Valley Lakes, which have undergone some of the most remarkable bouts of adaptive radiations in any system—aquatic or terrestrial. Cichlids also provide insights into how diversification can be accelerated by evolutionary innovations in a species, and by the climatic upheavals of the Pleistocene, even in regions far from the glaciers. Tragically, the current status of cichlids across these lakes also demonstrates how the legacy of such incredibly rapid radiations can be erased in "the blink of an eye" by human disturbance.

Galápagos finches and Hawaiian honeycreepers

Both Galápagos finches and Hawaiian honeycreepers descended from a single ancestral species of finch (Family Fringilidae) that colonized their respective archipelagos. Although the Galápagos finches have received the lion's share of attention from both scientists and the public—perhaps justifiably so given their impact on Darwin's seminal insights—the Hawaiian honeycreepers are the more remarkable in terms of their diversity. Whereas Darwin's finches include fourteen species from the archipelago and nearby Cocos Island, prior to colonization by Polynesians, Hawaiian honeycreepers included more than fifty species (this number may continue to grow as remains of recent extinctions are still being discovered in caves and other sub-fossil deposits across the Hawaiian Islands).

The Galápagos finches are all relatively modest in size and tend to be drab in color, while exhibiting substantial variation in key characteristics related to their diet, including the size and shape of their bills (Figure 14). These particular differences in their morphologies are reflected in an impressive breadth of habitats and habits, including species that exhibit different strategies for foraging on the ground or in the shrubs and trees for seeds and fruits. Other species of Galápagos finches have developed more specialized beak morphologies and behaviors: the woodpecker finch

1. *Geospiza magnirostris.* 2. *Geospiza fortis.*
3. *Geospiza parvula.* 4. *Certhidea olivasea.*

14. The Galápagos finches are relatively small and drab birds that differ most notably in the size and shapes of their bills, which correspond with differences in their diets and the habitats they occupy. Shown here are illustrations of four species from Charles Darwin's 1845 account, *Voyage of the Beagle* (1—large ground finch, 2—medium ground finch, 3—small tree finch, and 4—green warbler-finch).

utilizes cactus spines to poke into tree cavities for invertebrates, while the vampire ground finch pierces the skin of seabirds and iguanas to feed on their blood.

The morphological and dietary breadth and the degree of trophic specialization of Galápagos finches are, however, dwarfed by that of their Hawaiian counterparts (Figure 15). Although we may never know the full extent of the Hawaiian radiations, given that numerous species suffered extinctions before they could be "discovered" by scientists, the more than fifty described species of Hawaiian honeycreepers are much more diverse in terms of their body size, coloration, and the length and curvature of their bills. Bill morphology, in particular, was continually re-engineered by natural selection to more efficiently feed on a tremendous variety of resources including seeds, fruit, invertebrates, and nectar. The diversification of the nectarivores, in particular, was driven and likely accelerated by coevolution with another rapidly diversifying lineage—the Hawaiian lobeliads, an endemic plant lineage, discussed below.

The difference between the avian radiations in the Galápagos and Hawaiian Islands can be largely attributed to key differences in the geography of the two archipelagos (in particular, their isolation, size, and topography), and to differences in the ages of the islands or, of greater relevance, the age of the two lineages (i.e. the time since founding the archipelagos by the ancestral finches). Isolation is essential in adaptive radiations because it prevents gene flow among ancestral and descendant populations, which would otherwise swamp potential divergence. In their seminal 1967 monograph on island biogeography, E. O. Wilson and his colleague Robert H. MacArthur explained that there should be an optimal range of isolation for adaptive radiation (the "adaptive zone"), which varies as a function of the vagility (immigration powers) of the focal taxon. That is, to promote adaptive radiations, the islands in question must be within the dispersal range of a few individuals of select species, but well

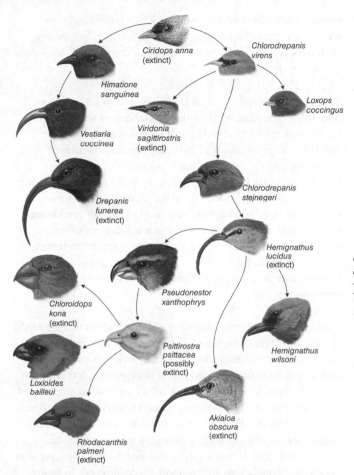

15. **Adaptive radiation of Hawaiian honeycreepers has far exceeded that of the Galápagos finches; here represented by just a small sample of the fifty or more honeycreepers which exhibit tremendous diversification in terms of their morphologies, diets, and habitats occupied. The numerous, now extinct forms represented in this small sample of Hawaiian honeycreepers attests to the vulnerability of island endemics in general.**

Within the figure:

Ciridops anna (extinct)

Chlorodrepanis virens

Himatione sanguinea

Vestiaria coccinea

Viridonia sagittirostris (extinct)

Loxops coccingus

Drepanis funerea (extinct)

Chlorodrepanis stejnegeri

Hemignathus lucidus (extinct)

Pseudonestor xanthophrys

Chloroidops kona (extinct)

Psittirostra psittacea (possibly extinct)

Hemignathus wilsoni

Loxioides bailleui

Rhodacanthis palmeri (extinct)

Akialoa obscura (extinct)

beyond that of most of its competitors; the latter condition ensures a variety of open niches that can be "invaded" during evolutionary and ecological divergence of the few lucky colonists to these remote islands.

So why were the radiations of Hawaiian honeycreepers more than three times greater than that of Galápagos finches? First, the Hawaiian Archipelago is much more isolated—roughly 3,600 km from the coasts of Asia (the presumed source of the lineage of finches that colonized these islands). In contrast, the Galápagos lay just some 950 km from the presumed origins of their finch lineage, South America. Two other measures of isolation may be even more relevant in terms of promoting adaptive radiations. The first is intra-archipelago isolation. Each island in these archipelagos has the potential to serve as a distinct evolutionary arena—"distinct" but not entirely independent, because inter-island colonizations were relatively frequent.

A second, more indirect measure of isolation may at first seem odd as it is a direct measure, not of distance, but of the size or area of each island. Evolutionary divergence may in fact occur among sites *within* an island provided that it is large enough to include natural barriers to dispersal and gene flow such as mountains and rivers. Because area and topography tend to co-vary, larger islands are more likely to include a greater range in elevation. Thus, the largest islands often include formidable mountains which, together with their streams, rivers, and ancient lava flows, serve to isolate local populations and promote diversification. To finally drop the other shoe, the Hawaiian Islands are indeed much larger than those of the Galápagos (maximum area 10,433 km^2 for the Big Island Hawaii, compared to 4,700 km^2 for the largest island of the Galápagos—Isabella), and the Hawaiian Islands are much more mountainous (maximum elevation 4,200 vs 1,700 m, respectively).

These two co-varying geographic factors—area and elevation—promote adaptive radiations in other ways as well. Larger islands with their greater range in elevation provide a greater diversity of habitats and potential niches to invade—enhancing opportunities for diversification over three spatial scales: on different mountains across the island; in disparate habitats on either side of the same mountain, as a result of the rain shadow effect; and at different elevations along each montane slope.

One last and often overlooked but essential role of island area is its effects on extinction, or in more positive terms, survival. Populations cannot diverge to any significant level if they suffer extinction within a few generations. If, on the other hand, they are lucky enough to colonize relatively large islands, then the ample resources and refugia (from severe storms, floods, and other potentially catastrophic events) afforded by those islands may provide the time required for ecological and evolutionary divergence among, as well as within, islands.

This brings us to one final factor especially relevant in the comparison of adaptive radiations of these two lineages of finches, one that builds on what we noted about the geological dynamics of these two archipelagoes in Chapter 2. The extant Hawaiian Islands are older than those of the Galápagos: the oldest extant Hawaiian island being Kaua'i, which emerged roughly 5.1 million years ago, while San Cristobal Island of the Galápagos emerged around 3.2 million years ago. But much more relevant than ages of the extant islands is, of course, the age of the lineages—that is, the time since colonization, or the number of generations for accumulation of island endemics across these archipelagos. And here, the difference is stark and compelling. The ancestral finches colonized the Galápagos around 0.5 million years ago (which may seem like a long time to most ecologists), but the ancestral, Eurasian rose finches that colonized the Hawaiian Islands established their first populations around 5.8 million years ago.

A careful check on the numbers here reveals, not a typo, but an apparent paradox—one that is resolved by considering the geological history of these islands. How is it that a 5.8 million year old lineage could colonize islands that didn't emerge for another 700,000 years? Remember that the Hawaiian Islands are just the most recent in a very long series of volcanic islands formed over a hotspot that drove the emergence of an immense, conveyor belt chain of islands extending some 6,000 km back to the Aleutian convergence zone and some 15 million years before the first ancestral finches arrived. Islands emerged over the hotspot, then joined the more ancient reaches of the chain of islands as they drifted on their tectonic plate away from the hotspot, successively slipping back beneath the waves as seamounts and guyots. Thus, it appears that the ancestral "Hawaiian" honeycreepers landed on a now submerged island many millennia before the extant islands of Hawaii emerged from the sea; persisting there and then island-hopping as new islands emerged along the chain from Kaua'i to its baby sister, the Big Island of Hawaii.

We can conduct similar comparisons of other lineages across many hundreds of other archipelagos and taxa, and the salient inferences would be the same. Adaptive radiations are strongly influenced by geographic factors—ecological and evolutionary divergence being most spectacular on those islands or archipelagos that are isolated, large, and mountainous, and for those lineages that are more ancient.

The above natural experiments attempted to "control" for differences among the focal species by selecting a very similar set of species (both passerine (perching) birds, and both descendants of finches). Yet species groups may differ in a number of ways that may strongly influence their propensities to undergo ecological and evolutionary divergence on islands (or other isolated systems). So let's now look at disparate groups of species occupying the same archipelago. First, we control for island differences by comparing the adaptive radiations of two lineages from the same archipelago—Hawaiian lobeliad plants and honeycreepers.

Hawaiian lobeliads and honeycreepers

The diversification of the Hawaiian lobeliads demonstrates one of the central properties of adaptive radiations—they are inherently *autocatalytic*. That is, they are driven in large part by interactions among the species themselves, which put a premium on species diversifying their niches and, in turn, their underlying phenotypic traits. At the microevolutionary (within species) level, this is wonderfully illustrated in the classic case of **ecological displacement** in the size and shape of bills among insular populations of Galápagos finches, with shifts among populations being more pronounced on islands where two incipient species co-occur. At the macroevolutionary level, divergence of one lineage can be reinforced and accelerated by that of another, in this case, an ecological mutualist.

Hawaiian lobeliads, which are flowering plants of the bellflower family (Campanulaceae), provide honeycreepers with food in the form of seeds and nectar and, in return, the honeycreepers serve as the plant's principal pollinators and seed dispersers. The close ecological associations of lobeliads and honeycreepers is showcased in the coevolution of flower structure and beak morphology in numerous pairings of host plant and honeycreeper (Figure 16). Such interdependence unfortunately brings its own liabilities, with the loss of either of these ecological mutualists having potentially devastating effects that cascade through biological assemblages.

Despite their close association across the same archipelago, the diversification of Hawaiian lobeliads has far surpassed that of their ecological associates, with over 125 species of these plants classified into six separate genera and occupying all islands and all

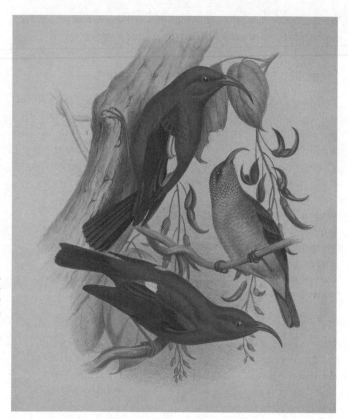

16. Hawaiian honeycreepers coevolved with one of their principal host plants—the Hawaiian lobeliads and, in mutualistic synergy, may have accelerated their respective adaptive radiations.

habitats—from the lowland, coastal forests to the highest vegetated zones of the Hawaiian Islands. Given that we are comparing forms from entirely separate kingdoms of life, the differences among relevant properties of lobeliads and honeycreepers are far too numerous to list here. But the much higher diversity of lobeliads may distill down to just two or three key factors. First, as plants, individuals can reproduce more rapidly than their vertebrate symbionts, so the turbines of evolution and speciation should spin much faster for the lobeliads. In addition, plants such as the lobeliads survive and divide the world up at a much finer scale (sometimes referred to as responding to their environment as more "coarse-grained"). Thus, what seems to represent just one habitat (e.g. a meadow) for birds may comprise a combination of smaller habitats for lobeliads.

Amplifying these differences in ecological and evolutionary propensities is a more than twofold difference in the age of these two lineages. The ancestor of Hawaiian lobeliads appears to have been a species of woody plant that colonized an antecedent, now submerged island approximately 13 million years ago, in contrast to 5.8 million years for the ancestral honeycreeper, and some 7 or more million years before the oldest of Hawaii's extant islands, Kaua'i, emerged. As Tom Givnish and his colleagues deduced from their extensive genetic and ecological research, evolutionary diversification of the ancestral lobeliads at first proceeded at a relatively modest rate, but then accelerated at about the same time that honeycreepers colonized the rising and still volcanically active island of Kaua'i. Once established on this emerging island, lobeliads diverged along what Givnish and colleagues describe as a set of "parallel, hierarchical radiations": *parallel* in that the temporal-spatial pattern of the divisions was repeated on each island; *hierarchical* in that each separate insular lineage underwent a similar succession of further splits on progressively finer spatial scales—first by principal habitat (open or forested), then across different montane slopes, then by elevational habitats along each slope (ranging from desert and coastal scrub to subalpine and alpine habitats), and finally at the finest scale by growth form (woody to herbaceous) and by habit (including diversification in flower type—apical or axillary; fruit type—fleshy or capsular; tube length; and inflorescence type). All this diversification was tightly linked to, and in large part driven by coevolution with honeycreepers, bees, butterflies, and other pollinators.

Madagascar's diverse and endemic lineages

All the features of the above examples of adaptive radiations across entire archipelagos are again evidenced in the remarkable assemblages of plants and animals inhabiting Madagascar.

Madagascar is an island with an area of 587,000 km². It stretches some 13 degrees latitude north to south, and lies roughly 400 km off the east coast of Africa (300 km during glacial periods of lowered sea levels). The diversity of Madagascar's endemic lineages of plants and animals is a product, at least in part, of its immense area and the resultant diversity of its habitats and potential niches. Madagascar is bisected by an expansive, central upland and myriad rivers and streams that flow toward the lowlands to further isolate native biotas into many separate, evolutionary arenas.

Madagascar also provides an illustrative case study of the various forms of isolation affecting evolutionary divergence. First, isolation has a temporal component. While sharing some biological affinities with India because of their ancient connection as part of Gondwana, Madagascar has been isolated from this and all other landmasses for roughly 90 million years. In addition to its time in isolation, and its contemporary geographic isolation from Africa and India, numerous endemic Malagasy lineages radiated in a fashion similar to the hierarchical pattern described by Givnish and his colleagues for the lobeliads of the Hawaiian Islands.

Perhaps better than any other system in the world, because Madagascar abounds in so many endemic lineages of plants and animals, its species assemblages present a compelling illustration of how isolation is a function, not only of distance and time, but of the dispersal capacities of the species as well. According to this concept of "functional isolation," regardless of how their ancestors arrived on the island, those with more limited dispersal powers and those with greater site fidelity (e.g. amphibians and non-volant (i.e. non-flying) mammals as opposed to bats) are likely to exhibit higher rates of diversification and higher levels of endemicity. This seems to be borne out by the inspection of available data on the endemicity of the diverse lineages now inhabiting Madagascar (Table 1).

Finally, the numerous lineages endemic to Madagascar provide another opportunity to assess the effects of lineage age on the magnitude of adaptive radiations. For simplicity, we can limit this comparison to one class of animals—the non-volant mammals, a

Table 1. Diversity and endemicity of Madagascar's native plants and animals. While many factors are involved, % endemicity tends to be highest for those species with more limited dispersal capacities (e.g. non-flying mammals vs bats and birds; snails and tiger beetles vs butterflies; flowering plants vs ferns—the latter with spores dispersed by winds)

Taxon/lineage	Arrival period (millions of years ago)	Number of species		% endemicity
		Total natives	Endemics	
Plants (flowering)	**25–15**	**13,000**	**11,600**	**89**
- 5 families with highest number of endemic species				
- Orchidacea		862	737	85
- Rubiacea		660	608	92
- Acanthaceae		512	476	93
- Euphorbiaceae		504	473	94
- Fabaceae		592	449	76
Pteridophytes (ferns, horestails, and lycophytes)	**23**	**586**	**265**	**45**

(continued)

Table 1. Continued

| Taxon/lineage | Arrival period (millions of years ago) | Number of species | | % endemicity |
		Total natives	Endemics	
Mammals		**155**	**144**	**93**
lemurs	~ 60	105	105	100
tenrecs	42–25	32	32	100
rodents	24–20	27	27	100
carnivores	26–19	10	10	100
bats	16–12	46	36	78
Birds	**45–26**	**313**	**181**	**58**
Reptiles		**384**	**367**	**96**
- freshwater turtles and tortoises	16 - 12	11	7	64
- lizards	60–40	240	230	96
- snakes	56–23	88	86	98
Amphibians (frogs)	**70–50**	**465**	**465**	**100**
Freshwater fishes	**164–145**	**164**	**97**	**59**
Land snails		**651**	**651**	**100**
Butterflies	**22–20**	**300**	**211**	**70**
Tiger beetles		**203**	**201**	**99**
Spiders (Araneae)	**16–12**	**459**	**390**	**85**

group whose ancestral colonizations span from 60 to 19 million years ago. The emergent pattern is consistent with our predictions. Species diversity is highest for those with the longest residence times—lemurs with over 100 endemic species in 60 million years; followed by tenrecs with roughly 30 species in 25 to 42 million years, and rodents with a similar diversity of endemics in roughly 20 to 24 million years.

In comparison to these omnivorous and herbivorous mammals, the diversity of mammalian carnivores is much less impressive, with just 10 endemic species for a lineage that colonized Madagascar some 19 to 26 million years ago. The more limited radiation of these carnivores may be largely attributed to their high energy demands, more specialized diets, broader home ranges, and longer generation times—all of this meaning the turbines of evolution spin much slower for these carnivores in comparison to lemurs and other herbivorous mammals endemic to Madagascar.

Over their long evolutionary history, lemurs have developed adaptations for an incredible diversity of trophic strategies and niches; all this supported by an equally impressive variety of behaviors, physiologies, and morphologies. Extant lemurs vary in body size from mouse lemurs of just 30 grams, to the indri which weighs around 9.5 kilograms. The actual range in body size of Madagascar's most defining mammals was, in fact, much greater than this. Madagascar was spared the Pleistocene extinctions that devastated most of the megafauna of other lands simply because it was not colonized by human populations until very recent times. But once colonized by humans, around 2,000–4,000 years ago, the pattern of collapse was the same as that following the arrival of ecologically significant humans across the globe—in this case, the selective loss of the seventeen largest lemurs, including the giant sloth lemur (*Archaeoindris fontoynontii*), which weighed nearly as much as an adult gorilla (body size estimates of the giant sloth lemur ranging from 160 to 240 kg; over 5,000 times the body weight of the still extant mouse lemurs).

Unfortunately, this story of megafaunal extinctions is all too common across the globe and across other taxa in Madagascar as well, including recent extinctions of other Malagasy giants—the giant fossa (a carnivore some 70% larger than the extant carnivores), two species of dwarfed hippos, and the elephant bird which, at over 3 m tall and weighing over 400 kg, was the largest (and not surprisingly, flightless) bird known to have existed.

Comparisons between radiations of these Malagasy mammals and those of its flying vertebrates (i.e.bats and birds) are consistent with the predicted effects of dispersal capacities on endemicity (Table 1). All species in the groups of non-volant mammals discussed above are endemic, while endemicity levels of Madagascar's bats and birds are substantially lower (78% and 58%, respectively). Endemicity is also very high for other, low-vagility vertebrates of Madagascar—100% for amphibians and 96% for reptiles. Among the invertebrates described in Table 1, endemicity is again highest for the species with more limited powers of dispersal—100% for land snails, 99% for tiger beetles, and 85% for spiders, versus 70% for butterflies. The two major groups of plants included in Table 1 also reflect the same pattern— ferns and their allies, which reproduce by spores that are readily and broadly dispersed by winds, demonstrating much lower levels of endemicity in comparison to flowering plants (45% versus 89% for ferns versus flowering plants, respectively).

The cichlids of Africa's Rift Valley Lakes

The lakes of east Africa's Rift Valley harbor some of the most diverse flocks of hyper-endemic species in the world. More than 1,400 species of cichlids (Family Cichlidae) inhabit the twelve largest lakes, together comprising some 60% of the global diversity of cichlids. The diversity of endemic cichlids is, consistent with patterns described for terrestrial islands, highest in the largest lakes—Victoria, Malawi, and Tanganyika, with over 300, 200, and 170 endemic species, respectively.

Comparisons between the radiations of these freshwater fish and the terrestrial lineages discussed earlier may at first glance seem confounded because we are comparing starkly different types of species and ecosystems. Yet research on adaptive radiations of cichlid fish across these lakes confirms many of the lessons garnered from the above descriptions of adaptive radiations in terrestrial systems. Analogous to the positive correlation between area and maximum elevation on land, lake area co-varies with maximum depth. Recalling our earlier discussion on plate tectonics, the Rift Valley is a tectonically active area where the continental plate of Africa is splitting apart, creating expansive depressions that over the past 30 million years filled in with rainfall and runoff to create the Great Lakes of this region.

In addition to these more long-term, geological dynamics, Africa's Great Lakes also were dynamic on a much shorter time scale. Each of the repeated climatic cycles of the Pleistocene repeatedly transformed the lakes—first isolating their deepest pools into separate lakes during glacial periods that brought drought to this region, and then refilling and reconnecting the glacial-period lakes during the much wetter interglacials. Cichlids thus experienced periods of diversification in isolation during glacial periods, followed by further diversification accelerated by ecological interactions when diverging species co-occurred during interglacial periods—all this repeated some twenty to twenty-five times during the past 2.6 million years.

In addition to these geological and geographic factors, the diversification rates of cichlids were strongly influenced by species traits as well. The cichlids of these lakes share a combination of characteristics that inhibit gene flow and promote evolutionary divergence and, at some early stage in their evolutionary history, they developed a particular morphological trait—an evolutionary innovation intricately involved in their incredible bouts of ecological and evolutionary divergence. In addition to exhibiting high habitat specificity—being limited to particular depths and restricted to various microhabitats by features such as sandy or rocky sediments—these cichlids have limited home ranges, are highly territorial, and exhibit strong **philopatry**: regardless of how far they disperse as fry, they return to their natal site to breed. So despite the open, three-dimensional nature of these aquatic ecosystems and the abilities of these fish to disperse to other parts of their lake, gene flow is spatially restricted often to just a few meters of their natal sites. As a result, these fish are ecologically as well as evolutionarily coarse-grained—undergoing divergence at relatively fine (local) scales within each lake.

The evolutionary innovation alluded to above is a second set of jaws (pharyngeal jaws), which join the typical, and more anterior, oral jaws. The principal function of jaws in vertebrates is to assist feeding—enabling capture, killing, mastication, and ingestion of prey by a hinge-like articulation of bone or cartilage. Possessing two sets of jaws allowed each pair to develop independently, freeing one up to "experiment" with alternative forms and functions—over time promoting enormous variation in the structures of jaws, teeth, and feeding habits of cichlids across the many and diverse habitats and microhabitats of these lakes (Figure 17).

The contributions of all these factors is, perhaps paradoxically, best demonstrated by the cichlids inhabiting one of the region's smallest and most recently formed lakes. Lake Nabugabo formed when a sand spit extended along the western shore of Lake Victoria (59,947 km^2) to form this tiny (22 km^2) lake. Despite being isolated for just 4,000 years, the lake has accumulated at least five endemic species of cichlids.

There is one final, sobering, but important lesson to draw from this fascinating system and incredibly diverse assemblage of species. Darwin's warning regarding the ecological naivety and fragility of the Galápagos's native wildlife is a foreboding but accurate one for all too many biotas evolving in isolation, including aquatic ones. Within just a few decades of its

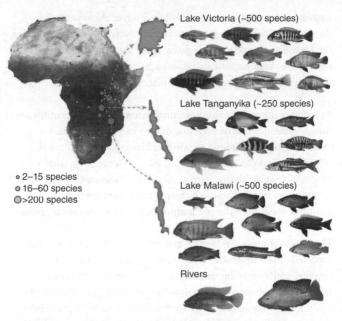

Lake Victoria (~500 species)

Lake Tanganyika (~250 species)

2–15 species
16–60 species
>200 species

Lake Malawi (~500 species)

Rivers

17. **The cichlids of Africa's Rift Valley Lakes represent a lineage that has undergone spectacular adaptive radiations among, as well as within, the lakes, much of this occurring during the climatic upheavals of the Pleistocene Epoch. Shown here on the right are images of just a tiny sample of the well over 1,000 species in this lineage.**

introduction to Lake Victoria, the Nile perch (*Lates niloticus*)—an intense competitor of cichlids when young and a voracious predator as it grows to reach nearly 2 m in length and over 240 kg—caused the extinctions of some 200 of the 300 cichlid species endemic to this lake.

Frontiers in research on adaptive radiations

The diverse assemblages showcased above represent just a small subset of the many compelling examples of adaptive radiations known, and probably a very small fraction of those yet to be discovered. Within the systems we have considered (in particular, the Hawaiian Islands, Madagascar, and the Rift Valley Lakes), many other taxa have undergone radiations perhaps as dramatic if not more so than the focal taxa discussed here.

Among the other renowned evolution arenas of diversification are the great menageries of vertebrates, invertebrates, and plants inhabiting coral reefs and isolated archipelagos throughout the tropical regions of the Atlantic and Pacific Oceans (e.g. *Anolis* lizards of the Caribbean, land snails of the Pacific archipelagos, and mammals of the Philippine Islands—the latter including an even greater diversity of endemic mammals per area than Madagascar).

These case studies of adaptive radiation, taken separately, and especially in comparisons across archipelagos and across taxonomically and functionally disparate groups of species, strongly affirm one of the central themes of biogeography—that evolution occurs across *space* as well as over time. In particular, the magnitude of adaptive radiations is strongly influenced by geographic variables, increasing with

- *isolation* (although there is likely an optimal range in isolation, the "adaptive zone," which varies depending on immigration powers of the focal species), and
- island *area*, which in turn co-varies with maximum elevation (or depth), and topographic (or bathymetric) diversity and environmental variation across the island (lake or ocean basin).

One key geographic dimension that may seem to have been overlooked in these natural experiments is latitude, but this was by design: not mine, but Nature's. All of the examples of adaptive radiations discussed earlier, and nearly all those reported, at least for extant biotas, are tropical. The intense solar radiation of the tropics not only powers the engines of photosynthesis—the ultimate source of food energy for nearly all forms of life on Earth—but it also drives the turbines of ecological and evolutionary diversification. The higher productivity and climatic stability of the tropics allows their species to be more specialized, thus allowing more niches and more species to be packed into an equivalent space in comparison to assemblages at higher latitudes. In addition, because the intense solar radiation of the tropics may accelerate mutation rates in many forms of life, and because maturation and reproduction rates tend to increase with environmental temperatures, evolutionary diversification occurs much faster in the tropics.

One other, autocatalytic feature of adaptive radiations deserves mention here. In addition to generating species with increasingly more specialized niches, adaptive radiation across the world's tropical islands and lakes may also include a trend toward reduced capacities or propensities for dispersal. This means that species at the more distal branches of the radiating lineages become effectively more isolated, with the spatial scale of evolutionary diversification then ratcheting down to finer levels: lineages diversifying first at the archipelago level, then among islands, across different watersheds within each island, etc. The radiations that led to the now roughly 800 species of Hawaiian drosophilids may be one of the most extreme examples of this spatial telescoping of species diversification, with extant species sometimes limited to distinct "kīpukas"—small tracts of land isolated by relatively recent lava flows as they reticulated down the mountain slopes across these tropical islands.

All rules of course have their exceptions, and perhaps the most glaring exception to the tropical cauldron "rule" of endemicity is the biota of Lake Baikal. Located far beyond the tropics at 53 degrees N latitude in Siberia, Lake Baikal is a global hotspot of endemicity for a variety of aquatic organisms, including over half of its 60 or so species of fish (these from three endemic families), over 500 endemic crustaceans, 80% of the lake's 150 species of snails, and the world's only exclusively freshwater species of seal—the Baikal seal (*Pusa siberica*).

Far from a *tropical* cauldron, but here again we have an exception that informs; in this case reinforcing our understanding of the forces driving evolutionary divergence and adaptive radiations. Three of the key promoters of biological diversity and endemicity—area, isolation, and antiquity—are at play in Lake Baikal. It is the world's largest (by volume) and deepest lake, and one of the most isolated and ancient bodies of fresh water—having formed in a rift valley that began to split open some 25 million years ago. Yet even the combined influence of Baikal's size and isolation have not fully compensated for its extra-tropical location. Although substantially smaller and less isolated than Baikal, the tropical Great Lakes of Africa's Rift Valley harbor far more diverse and endemic assemblages of most aquatic organisms. In short, while other factors strongly influence adaptive radiations, location is key and the tropics remain the pre-eminent evolutionary cauldrons for life on planet Earth.

Chapter 4 focuses on the rapidly advancing field of historical biogeography, which retraces and reconstructs the evolutionary histories of lineages over space and time. Chapter 5 then presents a regional to global-scale overview of patterns in species diversity across the principal geographic dimensions (area, isolation, latitude, elevation, and depth), before again focusing on the captivating products of adaptive radiations on islands, including the marvels and the perils of island life.

Chapter 4
Retracing evolution across space and time

The history of historical biogeography

The most defining and often visually compelling creations of biogeographers are maps, and they are of two distinct forms. Distributional or **systematic maps** are static in that they portray the geographic range or ranges of species during a particular time period, whereas **chorological maps** are dynamic in that they reconstruct the evolutionary development and geographic expansions of a focal lineage over time. The latter are of central interest to the sub-discipline of **historical biogeography**, and their origins can be traced back to the earliest naturalist/geologists of the 18th century.

Carolus Linnaeus (1707–78) dedicated his life to serving the Creator by describing and cataloging His divine menagerie of species. Linnaeus first described over 9,000 plant species and over 4,000 animal species, and also devised a system of organizing the burgeoning lists of species in the Creator's living museum—the system of binomial nomenclature we still use today. Linnaeus also developed one of the first chorological descriptions of the spread of all life forms from their primordial, paradisiacal homeland, and later from their mountain refuge following the biblical flood (the **Paradisiacal Mountain Theory**, Figure 18). Linnaeus' global-scale, historical reconstructions of the origins and spread

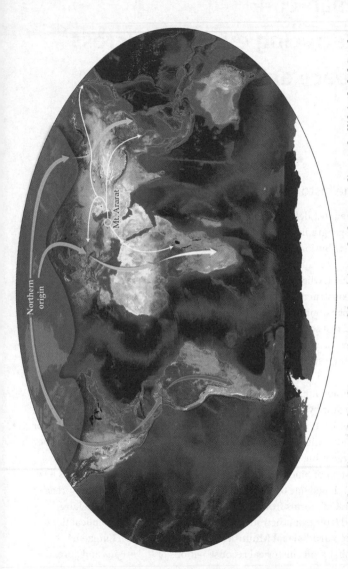

18. Two of the earliest, global-scale conceptions of the origins and subsequent dispersal of life included Carolus Linnaeus' Paradisiacal Mountain Hypothesis (from Mount Ararat, which during his time was believed to be the world's highest mountain), and Comte de Buffon's Northern Origins Hypothesis.

of Earth's life forms was limited by his gross underestimates of the actual global diversity of species and the true antiquity of the Earth (Linnaeus and his contemporaries thought our planet was just a few thousand years old).

Likely just as limiting was that Linnaeus was also bound by the doctrines of the fixity of Earth's continents, climates, and species. No one of that era could imagine that entire continents could drift across the globe like fragments of ice across a pond during a spring thaw. All of course accepted that climates changed seasonally, and catastrophically during times of divine intervention, but few scientists could fathom that the Earth itself experienced "seasons" of great winters that lasted many thousands of millennia. And the notion that, not just the Earth, but its species also were mutable—that they could "evolve" and create new species—not only was heresy, but inconceivable given what little was known about genetics during Linnaeus' time.

Later in the 18th century, in an attempt to explain why different regions of the globe are inhabited by different assemblages of species, Buffon proposed a model of the spread of life across the globe that relaxed two of the tenets on the fixity of the planet. His **Northern Origin Theory** was based on the revolutionary assertions that both Earth's climate and its species were mutable. The theory held that the ancestors of all of the life forms known to Linnaeus, Buffon, and their contemporaries once existed in the far north during a period of much warmer climatic conditions (Figure 18). He placed the initial, ancestral homeland of life high in the Arctic realm because its connections to both the Old and New Worlds would provide passage for species southwards to both as the Earth's climate later cooled to its current temperatures. Perhaps even more revolutionary (and heretical) in Buffon's Northern Origin Theory was that it asserted that species were mutable: they changed in their forms to adapt to the different environments they encountered as successive generations pushed southward through their new realms in either hemisphere.

The ultimate result was the final product of these range expansions across separate and environmentally dissimilar hemispheres—the distinct assemblage of species in the tropics of South America versus those of Africa—that is, the original and defining pattern of Buffon's Law.

Ernst Haeckel (1834–1919) was a contemporary of Alfred Russel Wallace, who, in addition to introducing or popularizing key biological concepts and terms including "ecology," "embryology," "phylum," "phylogeny," and "Protista," also argued cogently throughout his scientific career that any descriptions of the distributions and dynamics of life must be based on Darwin's theory of natural selection. Haeckel's map, shown in Figure 19, presents the geographic and evolutionary dynamics of our own species, from our paradisiac origins and then across land and sea as we diverged to form the regional populations recognized by Haeckel and his contemporaries at the dawn of the 20th century.

This small sample of classic contributions from the early history of historical biogeography demonstrates the interplay and interdependence between empirical knowledge and theory. Advancing theories of how life forms developed over space and time required a much better understanding of nature's patterns and, especially relevant here, the spatial and temporal dynamics of the Earth. It would take many more generations of innovative contributions from scientists across a range of fields before historical biogeographers could produce accurate reconstructions of the evolutionary dynamics of lineages over space and time.

Contemporary historical biogeography

While modern-day historical biogeography employs a burgeoning and diverse set of tools and approaches, it has one central goal: to create **phylogeographic** reconstructions of focal lineages— literally, tracts of the evolutionary divergence and radiations of a lineage over time and across the globe. At the heart of the

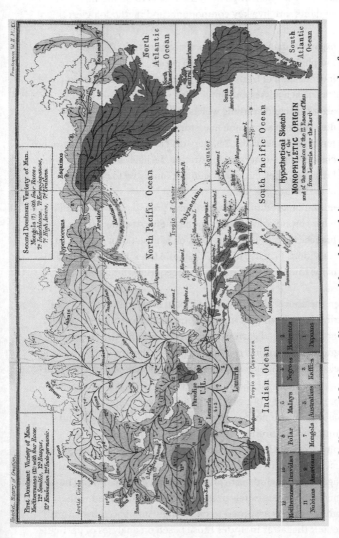

19. Ernst Haeckel's map of the human diaspora, although largely incorrect, is an early example of a chorological map, in this case illustrating Haeckel's ideas on the geographic origins and subsequent dispersals and divergence of early populations of our species.

methodology for constructing **phylogenies** (descriptions of evolutionary relationships between an ancestor and its descendants) are five key components and methodologies:

(1) expansive databases on characteristics or traits (genetic, morphological, etc.) of the focal species;

(2) identification of traits that are shared among those species;

(3) statistical analyses and application of a series of alternative algorithms to determine which of those traits are shared by descent from a common ancestor;

(4) further analyses using a battery of statistical programs to produce **cladograms**—diagrams illustrating the likely sequence of branching or evolutionary splitting (ancestral to most recent) of the lineage over time;

(5) utilization of additional algorithms, programs, and fossil and other chronological information to adjust the length of the branches to represent the duration of evolutionary splits and transitions to construct **phylogenetic trees**, that is, illustrations of the timing as well as branching sequences for the lineage.

Given the above, along with additional information on species distributions (both past and present), historical biogeographers can apply advanced cartographic programs and GIS to produce geographically explicit reconstructions of the evolutionary development of lineages across time and space—**geophylogenies**.

Before looking at examples of these phylogeographic visualizations, we need to flesh out some of the methods and assumptions for producing cladograms, phylogenies, and geophylogenies. Cladograms are constructed by comparing characteristics or traits of the component species in the focal lineage. These include morphological and genetic traits. The former often includes features most closely associated with breeding or traits assumed to be strongly influenced by natural selection; the latter includes those based on analyses of particular segments of DNA in the nucleus or in mitochondria of animals, or in the chloroplasts of plants.

The resultant, typically massive, databases describing trait states for each of the component species are then subjected to statistical analyses to identify those traits that are shared among species. The databases also include the trait states of a species that is related to, but outside the group of species under study. This "**outgroup**" provides a means to identify trait states that are ancient (ancestral) to the entire group ("**plesiomorphic**"), and those "derived" traits that arose during the evolutionary divergence of the focal lineage of species ("**apomorphic**"). It is the shared, derived traits—"**synapomorphies**"—that are most informative in constructing cladograms. The hypothesized sequence of branchings is estimated by assuming that the more synapomorphies shared by a pair of species, the more recent their split on the cladogram.

The branch lengths of cladograms are all of equal length because their intent is to portray the most likely *sequence* of branchings, not the actual or even relative time between each splitting event. Phylogenetic trees take this one step further by utilizing information on the ages when particular morphological traits appear in the fossil record, or on the rates of evolutionary divergence or mutation rates of DNA—so-called "molecular clocks." Once hypothetical phylogenies are developed, their alternative configurations can be assessed by a variety of independent means including, where geological reconstructions are available, comparing sequences and timing of the phylogenetic splits to those of tectonic splits of landmasses or ocean basins described by geologists in **area-cladograms**. An example of the latter would be area-cladograms reconstructing the sequence of island emergence, expansion, and submergence for the Hawaiian Islands—this being determined by geological analyses entirely independent of methods used to develop the species-cladograms and phylogenies (Figure 20).

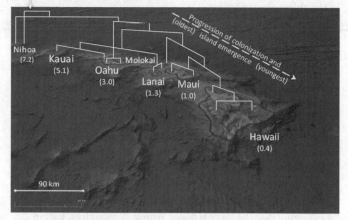

20. This phylogeny of Hawaiian katydids (genus *Banza*) seems generally concordant with the sequence of emergence for the islands (Hawaii being the youngest), indicating that dispersal and subsequent diversification of katydids largely followed the progression and emergence of islands (from ancient to more recent) along the Hawaiian chain as it developed over the past 7.2 million years.

Figure 20 shows an example of a geophylogeny. This geophylogeny of *Banza* katydids illustrates the arrival of the lineage onto a now remnant, pre-Hawaiian island (Nihoa) approximately 7.2 million years ago; that is, some 2 million years before Kaua'i (the oldest extant island) rose above the ocean's surface. The subsequent pattern and sequence of phylogeographic splits is consistent with what would be expected according to a **progression rule**—with descendant species colonizing each island as they formed in a conveyor belt-like fashion.

Geophylogenies are likely to be more complex for lineages evolving within archipelagos of more complex geological histories—that is, where the ages and geographic configuration of the islands do not follow a simple, linear progression. The geophylogenies of tortoises in the Galápagos are a case in point,

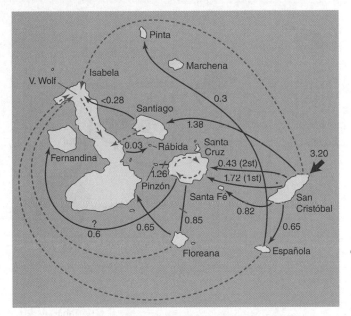

21. In contrast to the more simple progression patterns exhibited by many of Hawaii's lineages, this geophylogeny of Galápagos tortoises reflects the patterns of dispersal (natural and anthropogenic) of the species and the complex geological histories of the islands, which at times were joined and subsequently experienced splitting ("vicariance") of their lands and populations during alternating periods of volcanic activity. The thick black arrow indicates colonization of San Cristobal, the oldest island at around 3.2 million years ago, followed by subsequent natural dispersals (solid arrows) or human-assisted dispersals (dashed arrows). Islands that were previously joined and then split, due to submergence of a previous volcanic land bridge, are marked with a slash.

revealing occasional back migrations (from younger to older islands) and splitting of populations, or "**vicariance**," upon the submergence of earlier, volcanic land bridges between islands (Figure 21). Even for archipelagos with geological histories that do follow simple progression rules, geophylogenies may exhibit a complex, reticulating structure if the focal lineage

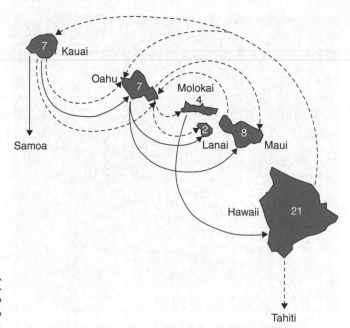

22. Even when island emergence follows a simple, regular sequence or progression rule, lineages comprising species with strong dispersal powers may exhibit relatively complex, reticulating geophylogenies caused by frequent back migrations of more derived forms to more ancient islands. Here shown are inter-island dispersals reconstructed for succineid amber snails of the Hawaiian Islands, with island ages increasing from the Big Island of Hawaii, in the southeast, along the chain toward the oldest island—Kaua'i in the northwest. Numbers for each island indicate the number of endemic species of this lineage of snails, and straight arrows pointing toward Samoa and Tahiti indicate dispersals and colonizations of those distant islands by Hawaiian snails.

comprises species with strong dispersal powers relative to the isolation among islands (for instance, the Hawaiian land snails—Figure 22).

The above geophylogeny for Hawaiian katydids is just one of many that are or will soon be available for this geologically and biologically dynamic archipelago, and even these visualizations comprise just a tiny fragment of the phylogeographic reconstructions being developed by the current generation of historical biogeographers. While each researcher may find some reconstructions more compelling than others, depending on our focus on particular taxa and types of ecosystems, it is likely that we all share a keen interest in the phylogeography of one lineage in particular: that of our own species and our closest, hominid relatives. I will come to the biogeographic dynamics of *Homo sapiens* in Chapter 7. Let us now consider how the new generation of statistical and technical tools is advancing the frontiers of phylogeography and simultaneously providing powerful means of studying biogeography's most fundamental pattern—the distinctiveness of place and the evolutionary divisions of the world.

Modern visualizations of Buffon's Law

The terminal branches of most phylogeographic reconstructions are extant species, which together define the biological distinctiveness of place—a concept central to biogeography and, in particular, its first and most fundamental pattern—Buffon's Law. Maps of biogeographic regions, such as Wallace's 1876 map of the world's zoogeographic regions (Figure 3), are in essence syntheses of distributional maps but, because their goal is to delineate the distinct arenas of evolutionary development, they are based on historical reconstructions of multiple lineages—that is, on an ensemble of phylogenies and, in particular, their terminal branches. So it should not be surprising that the same, or at least a similar set of approaches and tools used by historical biogeographers for constructing phylogenies and phylogeographies are also employed for delineating and describing biogeographic regions, the evolutionary divisions of the world. The goal of this line of historical biogeography is twofold: to delineate the regions of life (sometimes in a hierarchical scheme, e.g. Figure 23); and to provide descriptions of the biotic composition of each region

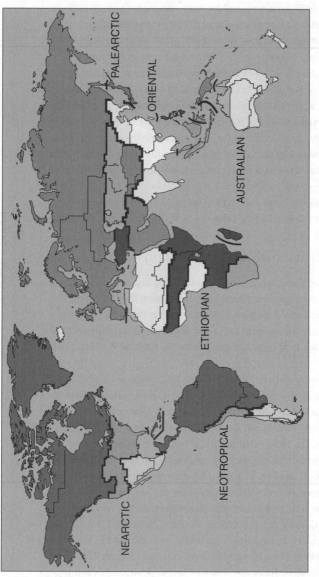

23. Similar to Alfred Russel Wallace's seminal map of the world's zoogeographical regions (Figure 3), other biogeographers have developed hierarchical maps of the world's biogeographic regions, in this case illustrating evolutionary divisions based on the statistical method of cluster analyses of similarity (shared taxa) among mammal assemblages at the species level.

along with estimates of its biological similarity to, or distinctiveness from, other regions (based on species shared between or endemic to particular regions, respectively).

Although phylogenies and biogeographic divisions of the world are developed by similar, or at least analogous approaches, their methods do differ in two fundamental ways. First, whereas the designation of distinct evolutionary units (i.e. species or subspecies) are the starting points of phylogenetic analyses, delineation of distinctive evolutionary arenas (biogeographic regions) is the end point or ultimate goal of regionalization analyses. Second, phylogenies are derived by analyzing similarity among species based on shared, derived characters (genetic or morphological traits). Biogeographic divisions of the world are also based on analyzing biotic similarity, but in this case the derived character states are not traits of the species, but the distinguishing characteristics of entire biotas—that is, endemic species. The hierarchy of biogeographic divisions from broad down to finer scales (e.g. regions, subregions, and provinces) are assigned based on similarity of assemblages across a parallel hierarchy of taxonomic levels (regions based on their distinctive, endemic families; subregions based on distinctive, endemic genera, etc.).

Wallace's seminal map of zoogeographic regions (Figure 3) illustrates his vision of the hierarchical nature of the biological divisions of the globe, and that we continue to use it today serves as a testament to his insights, drive, and massive accumulation of knowledge. Wallace's 1876 map, along with the voluminous descriptions of each region and subregion, were based on his exhaustive studies of distributional records for vertebrates—in particular mammals, which were strategically selected because they represented one of the better studied taxa at the time, and because they are relatively limited in dispersal powers (in comparison to birds or insects, for example) and are, therefore, more likely to be geographically restricted, that is, endemic to a particular region.

The methodological approaches, available data on geographic distributions, supplementary information on topographic and other physical barriers isolating biotas, and cartographic, GIS, and other tools available for visualizing the biogeographic divisions of the world have, of course, advanced tremendously over the many generations of scientists who built on Wallace's work. The ultimate goals, however, remain the same—delineating and describing the evolutionary divisions of the world. Global-scale biogeographic divisions are now available for a great variety of life forms ranging from mammals and other vertebrates, invertebrates, plants, fungi, and microbes across the terrestrial realm, to these and other taxa of marine ecosystems ranging from the coastal zones and pelagic surface waters of the oceans down to their abyssal depths (Figures 23 and 24).

One final note on the biogeographic divisions of the world. The very feature that stunned Buffon and his contemporaries, and eventually led to the revolutionary insights that would define the field of biogeography—the evolutionary distinctiveness of different regions—is now waning in the face of the geographic and ecological advance of one species: our own. Few taxa and regions across the globe have escaped the biotic homogenization caused by humanity. Regional biotas are becoming increasingly similar as a result of two pervasive, anthropogenic activities—extinctions of endemic species and species introductions. In fact, these two homogenizing effects of humanity are interrelated, with species introductions being one of the major causes of extinctions of endemic species.

Recall Gertrude Stein's lament over the loss in distinctiveness of place—that "*there is no there, there*." Tragically, this is becoming the sobering reality for the increasingly homogenized biosphere.

24. Although there are some noticeable similarities with Wallace's zoogeographic regions (Figure 3, which was based largely on mammals, birds, and reptiles), biogeographic regions developed for different groups of species exhibit some important differences; these likely reflecting differences in dispersal capacities and in the geographic origin and evolutionary histories of each focal biota. Shown here is a scheme for thirty-five biogeographic regions of the world based on distributions of land plants.

While we may not be suffering from the muted, *"Silent Spring"* that Rachel Carson warned us about in 1962, the monotonous cacophony of coquis (frogs native to Puerto Rico) and cicada in exotic lands as isolated as Hawaii now drown out the euphonious, more subtle calls of honeycreepers and other birds native to the islands.

Chapter 5
The geography of biological diversity

Biophilia, biodiversity, and the biogeographer's macroscope

As Darwin and Wallace taught us, natural selection has shaped the morphology, physiology, behavior, and ecology of all species, including our own. The knowledge of how the natural world varied from place to place—from one elevation to another along mountain slopes, from the coasts to the interiors of the continents, and outward from the shallow seas to the ocean depths—was essential for the survival of our ancestors. Natural selection for environmental knowledge was also manifested in, and reinforced by, an innate attraction to the natural world—**biophilia**—first discussed in general terms by the philosopher Eric Fromm in 1964, then applied by E. O. Wilson in 1984 to describe our genetically based connection with all of biological diversity. Biological diversity is another concept developed by Wilson in the 1980s which, as described in the opening to Chapter 1, is an all-encompassing term that includes variation in all characteristics of life from the chemical composition of cells to patterns in diversity and distinctiveness of entire communities of organisms.

Yet, as fascinating as we might find biological diversity, it is easy to be confounded by its seemingly limitless complexity. How are we to comprehend all of nature's diversity from the level of cells up through all taxonomic, biological, and ecological levels of organization? The answer, hopefully obvious by now, is to follow the lead of the great naturalists from von Humboldt, Darwin, and Wallace to Wilson and other contemporary scientists and apply what James Hemphil Brown describes as the biogeographer's **macroscope**.

Von Humboldt's classic macroscope was his visualization of the variation in plants, animals, and environmental conditions across the slopes of Mount Chimborazo, Ecuador. Darwin and Wallace's seminal insights on how evolution was driven by natural selection emerged from their powers to visualize incremental variation of populations of animals, not just over time, but across space as well (i.e. across islands of the Galápagos, and those of Indonesia, respectively). This, again, is the central mantra of biogeography: biological diversity is often rendered explicable when visualized across one or more of the principal geographic dimensions—area, isolation, latitude, elevation on land, and depth in the marine realm.

The meaning and measures of biological diversity

Throughout the over 200 years of the field of biogeography, its researchers have discovered some strikingly general patterns in biological diversity, and have advanced an equally intriguing set of explanations for the forces driving those patterns. Despite the many levels, qualitative features, and potential quantitative means of measuring biological diversity, the overwhelming majority of these studies have focused on just one or two relatively simple, but intuitively valuable measures—**species richness** and **endemicity**. Species richness is a simple count of the number of species in a particular area of interest (e.g. the number of fish in a pond, lake, or ocean basin). It is a direct, albeit simplistic expression of our innate value for the more complex. But our instinctive valuation of diversity is a bit more ecologically sophisticated than this, as it is also influenced by our apparently innate attraction to the rarest, most precious "gems" of the natural world.

A simple thought experiment should bear this out: given two assemblages with the same species richness—one comprising species common to most other ecosystems, and the other solely comprising endemics (so rare that they occur nowhere else), nearly all of us would be drawn to the latter assemblage because it has high endemicity. Beyond this instinctive attraction to the most rare, there clearly is a more pragmatic reason for valuing endemic species over the more broadly distributed (cosmopolitan) ones. If an endemic is lost from its assemblage, it disappears globally and the legacy of many thousands of generations of natural selection are irrevocably lost as well.

Geographic gradients across land and sea

The teaming diversity of the tropics. Johann Reinhold Forster was the ship's naturalist on James Cook's circumnavigational voyage from 1772 to 1775. He provided what appears to be the first scientific description of one of the most general and important, geographical-scale patterns in biological diversity—the **latitudinal gradient in species richness**. As Forster described for plants across islands of the southern hemisphere, and other scientists later generalized to plants and animals of all regions of the globe and nearly all types of ecosystems, species richness (or **species density**—species per standard sampling area) is highest in the tropics, and decreases as we move in either direction toward the poles.

Forster not only appears to have been the first to have clearly articulated the pattern, but he also provided a causal explanation, attributing it largely to the intense solar radiation of the tropics. While the now long list of hypotheses and explanations for the latitudinal gradient in species richness includes intense solar radiation as an ultimate factor driving diversification of tropical species, it may be even more fundamental than this—geometry. Among the contemporary hypotheses, the list of potential factors distills down to four distinct explanations, all ultimately associated with the spherical shape of our planet:

1. *Solar radiation*: Tropical ecosystems benefit from more intense solar radiation in the form of light and heat, the former driving higher rates of plant productivity, supporting more plants and, in turn, more animals—more herbivores, primary carnivores, etc. More intense solar radiation in the form of heat creates higher temperatures and accelerates growth rates, resulting in shorter generation times which, combined with higher mutation rates caused by more intense UV radiation, causes the turbines of evolution to spin faster in the tropics.

2. *Climatic stability*: The climatic conditions of the tropics are less variable over annual to longer time periods (e.g. through the glacial cycles of the Pleistocene), allowing their species to develop more specialized niches and, thus, more species can be packed into the same area or ecological space in tropical versus higher latitude ecosystems.

3. *Surface area*: Tropical landmasses and ocean basins are larger, providing a greater variety of resources, habitats, and potential niches, and supporting larger populations of plants and animals, which are thus less likely to suffer extinctions.

4. *The tropics are older*: In contrast to terrestrial or marine systems in higher latitudes, those in the tropics today tend to have been in the tropics for a longer period of time, giving more time to accumulate species by evolution or by immigration from other regions.

The final assertion on the antiquity of the tropics may not be intuitive, but, along with the other qualitatively distinct explanations above, all follow from geometry—that is, from the fact that the Earth is a sphere. Each of the conditions asserted in the above list and, consequently, the latitudinal gradient in species richness itself, would vanish if the Earth was flat or if, in some futuristic science odyssey, we designed a cylindrical, Second Earth. In such an alternative reality, contrary to each of the above points,

- the incidence of solar radiation would be the same across the globe;

- climatic conditions would only vary over time if the orientation of the cylinder changed; but at any one time, climates would vary little if at all across latitude;

- the middle latitudes (our "tropics")—on what essentially would be a rectangular map of our planet—would be no larger than zones at any other latitudes; and thus,

- if the plates were allowed to drift across our alternative Second Earth, they would spend just as much time in higher latitudinal zones as they would in the equal-sized middle zones.

Returning to reality, because Earth is a sphere, its surface intercepts the sun's rays most directly in the tropics (Point 1, above), which are delimited by the latitudes where the sun is directly overhead at least part of the year (from 23.5 degrees N to S). Because the Earth also spins on an axis that is tilted (whose orientation in the universe is toward the sun during our summer, away from the sun during our winter), we experience seasons, but they are much less apparent across the surfaces most perpendicular to the sun—the tropics (Point 2). The spherical nature and middle-bulge of our planet also means that the 47 degree-latitudinal breadth of the tropics circumscribes a greater area of landmasses and ocean basins (Point 3), which in turn means that even with the plates randomly drifting across the planet, they should spend a longer period of time in the larger zone—the tropics (Point 4).

One final point will reinforce one of our central lessons: *all patterns in biological variation across the globe result from factors influencing one or more of the three fundamental processes of biogeography—evolution, immigration, and extinction.* Put in this context, the reason why the tropics are teeming with species is that the factors described in Points 1–4 above combine to favor accumulation of species (through evolution and immigration), which far exceeds the loss of species (through extinctions).

Above and below sea-level. Two other gradients in species richness span a more localized geographic extent and, while not nearly as well studied as the latitudinal gradient, the emerging patterns suggest that they may be mirror images of each other. Although earlier observations claimed that diversity peaked near sea level and then decreased monotonically as we moved up along mountain slopes or down to the depths of the oceans, more recent studies reveal a qualitatively different pattern. For both the terrestrial and marine realm, diversity peaks at an intermediate elevation, or intermediate depth.

In attempting to explain this pattern, we must come back to the nature of the geographic template, which describes the sum total of multiple environmental factors as they co-vary across each of the principal geographic dimensions—in this case, elevation and depth. As we move up a mountain slope or down to the depths of the oceans, environmental conditions change in a complex, but qualitatively predictable manner—some becoming more favorable, others less so. For example, air pressure (including the partial pressures of oxygen and carbon dioxide) decreases as we move up in elevation.

One result of decreased air pressure is that air temperature also decreases (less pressure, less frequent collisions of air molecules); but because cold air cannot hold as much moisture, precipitation increases as we continue to climb up the mountain slope. In addition, whereas low-elevation habitats are more expansive and more interconnected (another matter of simple geometry), those at higher elevations are not only smaller, but also more isolated and, thus, their populations are less subject to gene flow and more likely to diverge genetically to form endemic species. Further, because they are geographically more isolated, populations occupying higher regions of the mountain slope are less likely to be affected by parasites, competitors, and predators (including humans) that spread easily across the more connected (and warmer), low-elevation habitats. Again, the patterns of co-variation in environmental features along a mountain slope are complex, but they distill down to some factors becoming more beneficial, others less so, with the optimal *combination* of these factors falling somewhere between the lowland and alpine zones.

Now let us return to the ocean's surface. As we move down in depth, a number of key environmental factors co-vary; again in a complex, but qualitatively predictable manner. The intensity of solar radiation, the primary source of energy for nearly all ecosystems, decreases with depth. Not only does this result in a rapid attenuation of light with depth, but temperature decreases with depth as well. On the other hand, deeper water zones do provide more stable conditions, being more isolated from storms and other disturbance events and, because all life forms eventually die and their remains sink, organic nutrients accumulate in deep water zones. Another positive feature of deep water zones is that they offer far more habitat than those in the surface waters.

This last point may be easily overlooked until we appreciate a key distinction between life in terrestrial versus marine realms. Nearly all terrestrial life forms live along the thin surface film of a few meters above and a few centimeters below ground. The marine realm, in contrast, truly is three-dimensional, with its inhabitants feeding, breeding, and evolving across the entire water column—the great majority of this lying well below the photic zone (roughly the upper 80 m where light is sufficient for photosynthesis). Thus, as we saw above for the elevational gradient, the optimal combination of environmental conditions for marine life should fall somewhere between the surface waters and the abyssal depths.

One caveat is in order: these two, mirror-image patterns—the mid-elevational and mid-depth peaks in species diversity, are not nearly as well documented as the latitudinal gradient in species diversity. Perhaps this reflects a genuine lack of consistency in these patterns among various taxa; for example, some peaking at mid-elevations, some in the lowlands, etc. It is just as likely, however, that these patterns are general but have emerged only recently because they required more detailed information on the distributions of species at much finer spatial scales (spanning just hundreds of meters versus thousands of kilometers for the latitudinal gradient in species diversity).

We are now benefiting from a resurgence in the interests of biogeographers, ecologists, and evolutionary biologists who are again following von Humboldt's lead and returning to mountain slopes as natural experiments in how biological communities respond to environmental variation across the geographic template. Simultaneous with these advances on land, our abilities to explore and conduct cutting-edge scientific research across Earth's most expansive, but logistically most challenging ecosystems—those of the marine realm—have advanced at a great pace over the past few decades, providing an increasingly comprehensive (yet still nascent) recognition of the diversity and geography of life across our "blue planet".

Species richness among islands

As we have already noted, islands have been especially instructive natural laboratories for some of the most transformative insights in biogeography, evolution, and ecology and, as we shall see in Chapter 7, island studies have played a central role in the more recent and critically important field of conservation biology. Let us turn again to the early explorers, and in particular Johan Reinhold Forster who, during his explorations of plant life across the southern oceans from 1772 to 1775, discovered two fundamental patterns in the biological diversity of islands, of which one was described some two centuries later as the closest thing to a rule in ecology—**the species–area relationship**.

In addition to showing that species richness of insular plants increased with island area, Forster also noted that more isolated islands had fewer species of plants (**the species–isolation relationship**), and their flora were more distinct than those of islands closer to the mainland. This latter observation is entirely consistent with our discussions on adaptive radiation in Chapter 3—that islands are often hotspots of diversification and endemicity provided that they are both large and isolated. This also helps to emphasize the importance of distinguishing between these two measures of biological diversity—diversity per se (species richness) and endemicity (the proportion of species that are endemic to that island).

In the two centuries since the explorations of Forster, biogeographers and ecologists have developed an increasingly refined description of the actual form of these patterns in insular diversity, and they have advanced an impressive body of theory to explain them. Regardless of the particular pattern, taxon, or type of islands or island-like ecosystems (e.g. lakes, caves, or fragmented forests), the ultimate explanations for these patterns must be based on first principles and, in particular, how area and isolation affect the fundamental biogeographic processes— immigration, extinction, and evolution.

Imagine cruising over the Earth's surface, setting the lens of our macroscope to register the species richness of entire ecosystems. Regardless of what type of systems we scan below—oceanic islands, lakes, coral reefs, patches of tropical rainforests, etc.—or which type of species we select on our macroscope's taxon channel, two nearly universal patterns emerge. The most general and widely documented of these island patterns is the species–area relationship: larger ecosystems harbor more species. But the pattern is a bit more nuanced than this. Our macroscope reveals that species richness first tends to increase rapidly as we scan from small to intermediate-sized ecosystems, but then the law of diminishing returns kicks in and the rate of increase slows as we view larger and larger systems.

Change our flight path so that we now scan islands along a route from the coastline outward, and the other general pattern emerges—the species–isolation relationship. Again, our macroscope reveals that this pattern is exhibited by nearly all types of species and ecosystems. Species richness decreases with increasing isolation, with the decline being steepest for the near islands and then attenuating and asymptotically approaching zero as we approach the most isolated ones.

For much of the early history of biogeography, explanations for these two patterns were based primarily on ecological factors. Larger islands could support more species because they intercepted more sunlight, which in turn resulted in higher carrying capacities for more plants and the other life forms they supported. Also, as we observed earlier, the nature of the geographic template is such that larger systems included a greater variety of environments (lowlands, mountains, lakes, rivers, marshes, etc.), which provide a greater variety of niches for species to inhabit.

Across our second flight path, in viewing islands of similar size we saw that those closer to the mainland typically are inhabited by more species. Just as species differ in their resource requirements and niches, they also differ in their dispersal capacities. Thus, near islands will be inhabited by most species from the mainland, but as we shift our focus to the more isolated islands, their assemblages become increasingly limited to include only the most powerful (or the luckiest) dispersers (bats, birds, and flying insects; rats and mice drifting on rafts of natural vegetation blown offshore by major storms; or the seeds and tiny snails that get caught in the mud on the feet of a migratory bird, or in its belly).

As biogeography, ecology, and evolutionary biology advanced into the middle decades of the 20th century, its researchers became increasingly frustrated over the absence of a general, unifying theory to explain diversity across all islands. Idiosyncratic hypotheses abounded: this island had this many species because it included these habitats, but not the other habitats required by different species; that island included all of these key habitats, but was too isolated to be reached by this subset of species; and so on. By the 1960s, biogeography, and in particular the burgeoning field of island biogeography, had reached a scientific crisis. Its practitioners were no longer complacent with the normal science of idiosyncratic and seemingly ad hoc explanations based on ecological traits of the particular islands and species under study. A scientific revolution was required to resolve the crisis and, as is often the case with transformative advances in the natural sciences, this would require a novel synthesis and collaboration between two of the leading ecologists and biogeographers of the 20th century—E. O. Wilson and Robert Helmer MacArthur.

Robert MacArthur was one of the leading mathematical ecologists of the middle decades of the 20th century, but also a gifted naturalist who produced some of the early classics on competition and coexistence of avian communities. Wilson was an equally gifted naturalist from his childhood, fascinated by insect communities, but also blessed with a remarkable power for inductive reasoning. You may remember from Chapter 1 how, after viewing his maps describing distributions of ants across the islands of Melanesia, Wilson developed his visionary theory of the ecological and evolutionary stages in development of insular lineages—from colonization of shoreline habitats by the ancestors to the ultimate extinction of their over-specialized descendants, now restricted to the island's interior habitats (Wilson's theory of the Taxon Cycle).

It may, therefore, seem a non sequitur that the seminal collaboration between these two highly accomplished naturalists would advance the field by turning a blind eye to species' differences and ecological interactions. And that is the beauty-in-simplicity that was at the heart of MacArthur and Wilson's new theory. It was species-neutral; that is, it was able to explain these two very general patterns in diversity among islands (the species–area and species–isolation relationships), while assuming nothing about any difference among species in terms of their resource requirements, niches, or dispersal capacities. Their theory, **the equilibrium theory of island biogeography**, was, however, based on first principles: patterns in species richness among islands resulting from the combined effects of the fundamental processes of extinction and immigration, which in turn are influenced by island area and isolation (regardless of the particular species).

Their theory, and in particular, its graphical model, focuses first on the dynamics of one island, which initially is assumed to be empty of species (Figure 25a). They then ask what happens to the rates of immigration and extinction as the island fills with species. Because they define immigration as the number of new species (those not already present on the island) arriving per time period, the rate of immigration to this island must decrease as the island fills up, because that leaves fewer species from the mainland that would be "new" to the island. Conversely, extinction rates should increase as species richness increases for the simple reason that, as the island fills, there are more and more species in residence that can go extinct. Given these very simple assumptions, and again assuming nothing about the characteristics of the species or the island, the two curves representing immigrations and extinctions must intersect (Figure 25a). At that point of intersection, the rate of species additions (via immigration) equals the rate of losses (via extinction), and this insular assemblage has reached a *stable, dynamic equilibrium*.

The equilibrium is termed "stable" because, if disturbed for some reason such that species richness drops below the equilibrium point, then immigrations will exceed extinctions—and species richness will rise. If, on the other hand, species richness rises above the equilibrium point, then extinctions will exceed immigrations and species richness will fall back toward equilibrium. The equilibrium is termed "dynamic" because, even though the focal island may maintain roughly the same number of species over time, the actual species inhabiting the island will change as new immigrants replace those that go extinct.

An elegantly simple model, indeed; but so far it says nothing about area and isolation. To fully develop their species-neutral

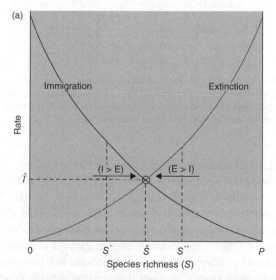

25. **Robert H. MacArthur and E. O. Wilson's graphical model of their equilibrium theory of island biogeography was developed in two parts: first (a) illustrating how the two fundamental processes of immigration and extinction should vary as an originally empty island fills up with species, and (b) illustrating how the rates of these processes should vary for islands that differ with respect to their area and degree of isolation—ultimately explaining the species–area and species–isolation relationships.**

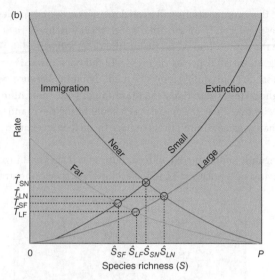

25b. Illustrating how the rates of two fundamental processes of immigration and extinction should vary for islands that differ with respect to their area and degree of isolation—ultimately explaining the species–area and species–isolation relationships.

explanation for the species–area and species–isolation relationships, MacArthur and Wilson superimposed one additional layer of complexity onto their single island model—basically by asking how the extinction and immigration curves would differ among islands of different size and degree of isolation (Figure 25b). Again, the answer is elegantly intuitive. Extinction rates should be lower on larger islands because they have more resources; thus, as area increases the intersection point, or equilibrium level of species, should shift to the right (higher species richness on larger islands). Immigration rates, on the other hand, should vary with isolation—decreasing for the more isolated islands; thus, as isolation increases the equilibrium number of species should shift to the left (lower species richness on more isolated islands).

MacArthur and Wilson's 1967 monograph describing their equilibrium theory quickly became a citation classic, revolutionizing how we studied the dynamics of islands. Throughout the remaining decades of the 20th century, a generation of scientists would test the assumptions of their species-neutral model, assess its generality over a broad range of ecosystems and taxa, and explore its potential applications for conserving species on oceanic islands and across other, island-like systems as well. By the new millennium, however, the pendulum began to swing back as a new generation of scientists began to question the simplistic assumptions of MacArthur and Wilson's theory: its assumptions of a long-term, stable equilibrium; of species equivalence; and its absence of ecology. Instead we began to search for alternative theories that could explain a broader variety of patterns in **community assembly** (the non-random accumulation of non-equivalent species).

Ironically, one of the earliest challenges to the equilibrium theory and its simplistic assumptions would come from Wilson himself, along with one of his most distinguished students and leading ecologists of that era—Daniel Simberloff. For Simberloff's dissertation, the team decided to test the central assumptions of the theory—that an empty island would accumulate species and reach a stable equilibrium, but the equilibrium would remain dynamic as new species replaced those that went extinct. They devised an ambitious and ingenious field experiment on a set of tiny islands in the Florida Keys. After clearing the islands of all invertebrates with methyl-bromide, they meticulously surveyed the islands over the next two years, developing detailed lists of each species that arrived or suffered extinction. At first, their results seemed entirely consistent with the basic tenets of the theory. Species richness rapidly increased and then slowed to approach an apparent equilibrium on each island, while species replacements continued. To their surprise, however, Simberloff and Wilson witnessed not just one, but two and possibly three qualitatively distinct equilibria for each island (Figure 26).

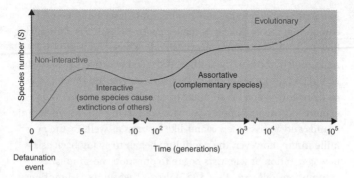

26. A hypothesized time series of equilibria suggested by Daniel Simberloff and E. O. Wilson's classical, manipulative field experiments on defaunation and recolonization by invertebrates of tiny islands in the Florida Keys.

After reaching an initial (what they termed "non-interactive") equilibrium, species richness declined somewhat to level off at a lower, "interactive" equilibrium—presumably resulting from negative interactions (competition and predation) among species, which drove some to extinction. Over time, however, species richness again rose toward what the two scientists speculated might be a third equilibrium—an "assortative" equilibrium resulting from many generations of ecological sorting for complementary (more mutualistic) suites of species. Although far beyond the spatial and temporal scope of Simberloff and Wilson's field experiments, we might envision a final, evolutionary phase of the species–time curve of Figure 26, provided that the islands are large enough and the "experiment" runs long enough to enable in situ speciation. Given the logistic challenges of conducting rigorous studies at such broad spatial and temporal scales, we can only speculate as to whether species richness on very large and very old islands would continue to rise, or ultimately achieve a fourth ("evolutionary") equilibrium.

Simberloff and Wilson's classic field experiments provided insights that went far beyond initially appearing to confirm the basic tenets of MacArthur and Wilson's theory. The integration (or reintegration) of species differences, ecological interactions, long-term dynamics (including non-equilibrial assemblages) and evolution back into island biogeography theory proved seminal to a generation of empirical and theoretical advances in island biogeography. We can find beauty in complexity as well as simplicity—provided we can strategically devise the appropriate macroscope to visualize those patterns and develop more integrative theories to explain them. Simberloff and Wilson's field experiments and conceptual models, and a legion of others developed by their colleagues and subsequent generations of scientists, demonstrated that the species–area and species–isolation relationships may be more complex than initially conceived—varying, albeit in a predictable manner, among the world's oceanic islands, and other island-like systems as well.

The heuristic value of studying the species–area relationship goes far beyond a so-called curve-fitting exercise. Our explorations of this pattern, and in particular its causal nature, tell us of something fundamental about how nature works and how biological communities of different size are assembled. But to fully appreciate patterns in biological diversity among islands, we need to better understand the true nature of the species–area relationship (Figure 27). Rather than following the simple, monotonic pattern described above, the species–area relationship appears to be protean, taking on different forms depending on the particular range in area we consider (in Greek mythology, Proteus was a sea god capable of taking on many different forms). For very small islands, species richness remains low and appears to vary independently of island area. For larger islands, species richness traces the more conventional, monotonic trajectory with increasing area described earlier. For very large islands, however, the relationship appears to enter a second phase of rapid increase (in this case with little suggestion of an asymptote). This, admittedly, is far from the simple pattern we described earlier as

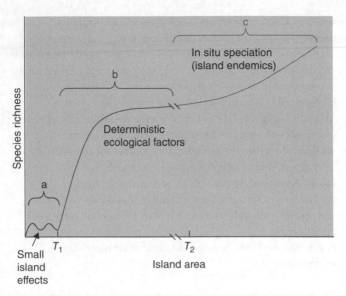

27. **The species–area relationship may be more complex than originally conceived—exhibiting three different, scale-dependent phases ranging from (1) where species richness tends to vary independently of area on the very small islands, (2) increasing with area consistent with the conventional form of the species–area relationship for islands of intermediate size, (3) but then exhibiting another phase of accelerated increase in richness with island area for islands large enough to provide the conditions necessary for in situ (i.e. within island) speciation.**

"the closest thing to a rule in ecology"; but here again we can find beauty in complexity. The three phases of this expanded species–area relationship provide a valuable lesson concerning the scale-dependent forces assembling insular communities, both great and small.

Small islands are very open systems—being most strongly influenced by external forces (storms, floods, etc.) rather than by internal processes, such as ecological interactions among the few species that may gain a foothold on these tiny specks of land. Eventually, however, our macroscope widens to scan beyond this stochastic, **small island effect** phase to include islands that are large enough to provide adequate resources and refugia for buffering their inhabitants against the stochastically varying, external forces. At this point we enter the prototypic range of the species–area relationship: an **ecological phase** driven by the combined effects of immigrations, extinctions, and species interactions.

Finally, our macroscope expands to its maximum extent to capture the final phase of the species–area relationship—the **evolutionary phase,** where islands are so large that they themselves become evolutionary arenas for speciation and adaptive radiations as discussed in Chapter 3. Only on these very large islands do we see the mountain ranges, rivers, and other dispersal barriers that provide the within-island isolation necessary for the formation of endemic species among different parts of the island. Thus, across these three scales of island area, our macroscope has captured the manifestations of all three fundamental biogeographic processes—immigrations and extinctions operating stochastically across very small islands (the stochastic, small-island phase); immigration and extinctions (likely mediated by species interactions) operating in a more deterministic manner on the larger islands (the ecological phase); and all three processes—immigration and extinction now joined by speciation (again, all strongly influenced by interspecific interactions) on the very large islands (the evolutionary phase).

To capture the full complexity and beauty of biological diversity across the world's oceanic islands, we need to readjust our macroscope to expand again—in this case not across space, but over time. The revolutionary insights of Alfred Wegener and the generations of geologists that followed was that the Earth's geographic template is dynamic, and that this involves not just the drifting, collision, and rifting of the continents, but the formation, emergence, and destruction (subduction) of the ocean floor, including the dynamic lives of oceanic islands. So if we and our macroscopes could go back in time and if we were patient enough, we could observe the life cycle of an island from its volcanic emergence from the ocean floor to eventually crown above the surface of the sea, expanding in area and topographic complexity until this portion of the oceanic plate drifts beyond the hotspot deep in the mantle below, with the island then retreating back toward the water's surface and the ocean floor as erosion and subduction take their toll.

Enter Robert J. Whittaker, whose dissertation research focused on the classic example of island dynamism—the cataclysmic explosion and defaunation of the Krakatau Islands in 1883, and their subsequent recolonization by plants and animals over the next century. For the next three decades following his dissertation (likely with the images of island dynamics simmering at the back of his mind), Whittaker continued to provide insights advancing the field of island biogeography, and biogeography in general. Then, during the early decades of the new millennium, Whittaker and his colleagues formulated their **general dynamic theory of island biogeography**, which extrapolated from MacArthur and Wilson's equilibrium theory, in synergy with plate tectonic theory, to explain how the geological and physical features of an island undergo a cycle of changes from its first emergence above, to final submergence beneath the waves. These geophysical dynamics, in turn, drive the fundamental processes of immigration, extinction, and speciation (and, therefore, species richness) to rise and fall in a highly predictable sequence of patterns during the life cycle of an oceanic island (Figure 28).

Taken together, Whittaker's general dynamic theory, along with the research on adaptive radiations across volcanic islands such as

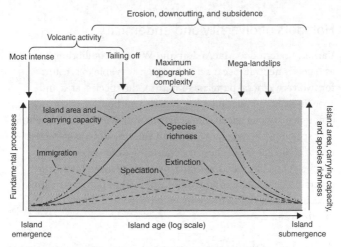

Erosion, downcutting, and subsidence

Volcanic activity

Most intense | Tailing off | Maximum topographic complexity | Mega-landslips

Island emergence | Island age (log scale) | Island submergence

Fundamental processes

Island area, carrying capacity, and species richness

Island area and carrying capacity
Immigration
Speciation
Species richness
Extinction

28. **The general dynamic model of Robert J. Whittaker and his colleagues, which depicts regular changes in the fundamental biogeographical processes (immigration, extinction, and evolution) and the total number of species—all this driven by dynamics in the physical characteristics of a volcanic, oceanic island during its life span.**

those of the Galápagos and Hawaiian Archipelagos, serve to re-emphasize the central lesson that evolution occurs across space as well as over time. Perhaps just as important, they explain why it may be difficult if not impossible to assess the independent influence of each of these two dimensions (spatial versus temporal) without explicitly considering the other. By melding MacArthur and Wilson's equilibrium theory, which dealt with dynamic biotas inhabiting geophysically static islands, with plate tectonic theory as applied to the geological dynamics of islands, Whittaker and his colleagues provided a powerful, spatio-temporal macroscope and an emerging body of theory capable of explaining one of the most important patterns in biological diversity, the species–area relationship, and enabling the discovery some new ones as well.

Hotspots of diversity and endemicity

Earlier, I referred to MacArthur and Wilson's equilibrium model as "species neutral" because, although it was able to account for two very general patterns in nature, the species–area and species–isolation relationships, it did so without invoking any differences among species. MacArthur and Wilson knew of course that species were not equivalent, but their model was so robust that it could ignore all these confounding differences among species yet still explain the two general patterns in biological diversity. Then again, many if not most other patterns in and measures of biological diversity go far beyond a simple numbers game.

The "names" of the species or, more to the point, their distinctive morphologies, physiologies, behaviors, ecological interactions, and evolutionary histories, are often the very features we most value as scientists. Accordingly, decades of biological surveys and syntheses of the vast accumulations of data they produced have enabled the identification of global hotspots of biological diversity and endemicity.

The results are global-scale macroscopes, or maps of biological diversity such as those shown in Figure 29a–c, which identify hotspots of diversity and endemicity for flowering plants and mammals. While the locations or intensities of hotspots may differ depending on the particular taxon of interest, the general, salient patterns are consistent with our understanding of adaptive radiations and the geography of evolution. Hotspots of diversity and, in particular, those of endemicity, are concentrated in tropical regions, and especially in large, topographically complex, and isolated systems—those large enough to provide a diversity of resources at levels requisite for maintaining populations long enough to allow evolutionary divergence, and those of sufficient

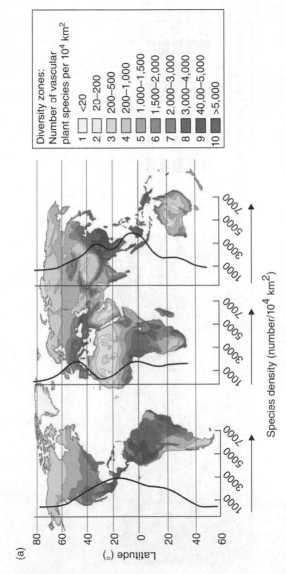

29a. Global hotspots of diversity (measured as species/standardized area) for vascular plants.

Latitude (°)

Species density (number/10⁴ km²)

Diversity zones:
Number of vascular plant species per 10⁴ km²

1 | <20
2 | 20–200
3 | 200–500
4 | 200–1,000
5 | 1,000–1,500
6 | 1,500–2,000
7 | 2,000–3,000
8 | 3,000–4,000
9 | 40,00–5,000
10 | >5,000

113

Species per ~22,300 km² hexagonal cell

Terrestrial

1
37
75

131
184
274

Marine

1
10
19

25
30
41

29b. Global hotspots of diversity (measured as species/standardized area) for mammals.

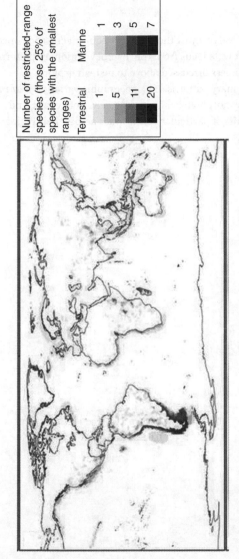

Number of restricted-range species (those 25% of species with the smallest ranges)	
Terrestrial	Marine
1	1
5	3
11	5
20	7

29c. Compare the previous map of mammalian richness, per se, to that illustrating hotspots of endemicity for mammals (based on the numbers of species with restricted ranges).

isolation to prevent the genetically homogenizing effects of gene flow.

In Chapter 6, we readjust the settings of our macroscope once again to narrow its focus from the macroevolutionary patterns in species richness discussed above to instead explore microevolutionary patterns of biogeographic variation *within* species—how morphological, physiological, behavioral, and ecological traits of particular species vary across their regional populations.

Chapter 6
Macroecology and the geography of micro-evolution

Macroecology—emergent patterns in the geography of life

It is time to more fully describe the conceptual foundations of macroscopes, which provide multi-scale windows into the complex structure of the natural world. As I mentioned earlier, the term "macroscope" was first introduced during the latter decades of the 20th century by James Hemphil Brown. Jim is a singular scientist, who somehow blends an encyclopedic memory with an insatiable curiosity for all scientific phenomena, and the creativity and imagination to visualize patterns across multiple scales of space, time, and biological organization. His research interests range from the physical laws engineering the micro-structure of cells, to the many influences of an individual's body size on its physiology, life history, and ecological traits, and on to the forces structuring entire biological communities, regional biotas, and global-scale patterns in biological diversity. Who then would be more capable of not only shepherding the modern renaissance of biogeography but also advancing the field of **macroecology**—a conceptual framework for visualizing multi-scale patterns across all realms of the natural sciences?

One of the central tenets of macroecology is that traits characterizing one level of organization (e.g. traits of individuals) form patterns that become emergent at higher, or broader levels (e.g. at species or community levels). A characteristically elegant macroscope of Brown's will illustrate this "emergence" of patterns across scales of space, time, or biological organization. The macroscope in Figure 30 illustrates the relationship between body size (an individual-level trait) and geographic range size (a species-level trait), which yields an emergent pattern characterized by salient features which Brown and his colleagues term "constraint lines." The recognition and emphasis on constraint lines represents a novel advance in that, rather than limiting our inferences to those addressing central tendencies through a cloud of data points, macroecologists teach us to also consider the margins of the clouds and the forces constraining those margins (here, bivariate data representing combinations of body size and range size for different species).

The macroscope of Figure 30 reveals three constraint lines: a vertical line interpreted as a physiological limit to the minimum body size of a mammal (around 2 grams); a horizontal line indicating a geographic limit to the maximum size of a mammal's geographic range (about half of the continent's size); and a third, diagonal constraint line that has special relevance for conservation biology as well as biogeography. Extinction (a population- and species-level trait) is most likely to occur, not simply for species with small geographic ranges per se, but for those species whose ranges are (or were) small *relative to* their body size. Thus, in this one macroscope, we can visualize the importance and interdependence of morphology, physiology, biogeography, and extinction (or endangerment).

Let us now explore a second, equally creative macroscope of Brown and his colleague Brian Maurer, in this case serving to illustrate the relationships between size and the shape of geographic ranges, and how these **areographic** features are influenced by continental-scale topography and by glacial cycles of the Pleistocene. Areography is a set of analyses and visualizations

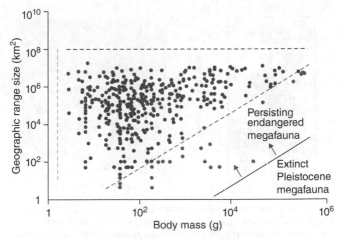

30. **A macroscope illustrating the relationships between body size and geographic range size of mammal species of North America, also illustrating three constraint lines (dashed lines, explained in the text), and a solid line that approximates the position of the diagonal constraint line prior to extinctions of the Pleistocene megafauna.**

first developed by Eduardo H. Rapoport in the 1970s and 1980s to explore the relationships between the internal structure of geographic ranges (e.g. how population density varies across the range of a species) and their external features (size and shape). Brown and Maurer devised an elegantly simple method of measuring the sizes and shapes of geographic ranges of species by first measuring their north–south (N–S) and east–west (E–W) extents, and then plotting those values on a graph such as that of Figure 31a. Roughly circular (or symmetrical) ranges fall along the line of equality, with range size increasing from the origin to the upper right region of the graph (Figure 31b).

After plotting these data for birds of North America, Brown and Maurer discovered that, as they considered species with larger ranges, the shapes of those ranges were transformed from being roughly symmetrical for species with the smallest ranges, to being

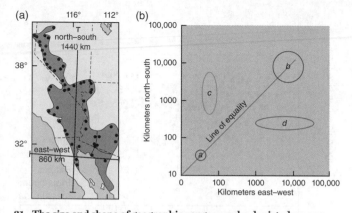

31. The size and shape of geographic ranges can be depicted on one graph by (a) measuring and then plotting the north–south and east–west extents of the ranges, and then plotting these values for each species on a graph such as that in (b). Points on the line of equality in part (b) denote species with symmetrical ranges, while those above or below the line represent species whose ranges are stretched N–S or E–W, respectively.

stretched N–S for those with intermediate-sized ranges, then once again symmetrical for species with larger ranges, and finally where the ranges became stretched E–W for species with the largest ranges (Figure 31c). This pattern of transformations in range size and shape appears to reflect two scale-dependent features of the geographic template. Starting at the origin, where ranges have to be symmetrical, larger (intermediate-sized) ranges are molded by principal topographic barriers, which happen to run N–S in North America. Species with larger ranges must be able to disperse beyond these barriers, but now their ranges are limited by climatic zones, which are stretched E–W (i.e. across latitudes).

Brown and Maurer went on to test their explanation for this emergent pattern in ranges of North American birds by applying the same macroscope, but this time to birds of Europe (Figure 31d), where the principal topographic barriers (the Pyrenees, Alps, and

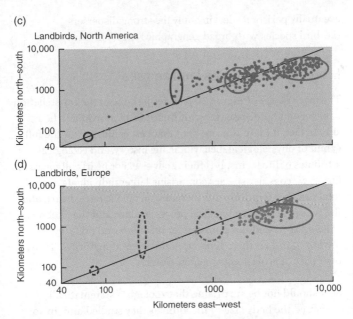

(c)

Landbirds, North America

(d)

Landbirds, Europe

Kilometers north–south

Kilometers east–west

Macroecology and the geography of micro-evolution

31. **Macroscopes of the relationships between size and shapes of geographic ranges of birds in (c) North America and (d) in Europe are similar, except for the paucity of bird species with small geographic ranges in Europe.**

Caucasus Mountains, and the Mediterranean Sea) run E–W (i.e. not N–S as for North American barriers to dispersal). The key distinction between emergent patterns of these two macroscopes is the absence of birds with small to intermediate-sized geographic ranges in Europe. The ultimate cause for these differences can be traced back to the influence of glacial cycles of the Pleistocene on the shifting ranges of these birds. In North America, even those birds with relatively limited dispersal powers (and, therefore, with small- to intermediate-sized ranges) were able to adapt by shifting their ranges N–S with each glacial cycle. Their counterparts in Europe, however, were blocked by formidable mountain ranges and by the Mediterranean Sea, and they

eventually perished—leaving only the strong dispersers
(i.e. bird species with broad geographic ranges) in Europe.

Ecogeography across land and sea

The macroscope of Figure 30 features just one of the constellation
of factors and processes strongly linked to an individual's body
size. In fact, if there was one trait that best characterizes the full
gamut of physiological, behavioral, life history, and ecological
attributes of life forms, it is their body size. For nearly all forms of
life, as we consider species comprising larger individuals, their
total resource requirements increase; metabolic rates, heart, and
the pace of other physiological processes slow; and the likelihood
of being hosts and not parasites, or predators and not prey,
increases, as does their generation time (age at maturity). Thus,
their pace of evolutionary change slows.

So it should not be surprising that zoologists systematically
measured the body size of the animals they studied and, in so
doing, often discovered some striking patterns in the
morphology and geography of life forms. Such patterns are
described as morpho-geographic, or, more commonly, as
ecogeographic patterns. Among them are the differences in
body size between insular animals and that of their mainland
ancestors, and a select number of ecogeographic gradients
discovered for vertebrates across the continental and marine
realms. After Darwin and Wallace published their theory of
evolution by natural selection, their colleagues set out to discover
how selective regimes molded regional biotas and their
morphological variation from place to place.

In 1847 Carl Bergmann articulated a latitudinal gradient in body
size of "warm-blooded" animals (mammals and birds) that seemed
so general that it became known as "Bergmann's rule": the
body size of individuals of the same, or closely related, species
tends to increase as we move from the tropics to the poles or,

equivalently, from regions with warm to those with colder climates. In the marine realm, a corollary of this body size gradient was reported by David Starr Jordan in 1891. According to "Jordan's rule," the number of vertebrae and, therefore, the body length of bony fish tend to increase from tropical to colder reaches of the oceans. Both of these related, body size gradients can be explained on the basis of natural selection across different temperature regimes. Larger animals have more energy stores and—especially important for birds and mammals that maintain their body temperatures by their own metabolic heat—more insulation (in the form of feathers, fur, and fat). Thus, larger animals should be more "fit" (more likely to survive and reproduce) in regions of cold temperatures and limited food supplies.

These geographic gradients in body size may also have an alternative, ecological explanation related to the latitudinal gradients in species richness we considered earlier. Remember that, as we move from the Equator to the Poles, not only do climates get colder, but the diversity of biological communities (i.e. of predators, competitors, etc.) decreases. In species-rich communities such as tropical rainforests, competition often causes species to differ in body size, in turn reducing their niche overlap and promoting coexistence—the process known as ecological displacement. Along the gradient from the tropics to the poles, because species are increasingly released from these ecological pressures, smaller species often increase in body size in the absence of their larger competitors (**ecological release**). So the body size trends of Bergmann's and Jordan's rules may be attributed to thermoregulatory adaptations to the cold, to ecological release in species-poor (high latitude) communities, or both.

In 1878, Joel Asaph Allen discovered another ecogeographic trend in birds and mammals that appears driven by natural selection for optimal phenotypes under different climatic regimes. As we move from the tropics to the poles, the appendage length of birds and mammals decreases, thus reducing the surface area across which they would otherwise lose heat to cold environments. In contrast, Constantin Gloger described an ecogeographic pattern in 1883 that appears to be independent of thermoregulatory constraints, and more closely associated with ecological pressures—in particular, those between predators and prey. Pelage (fur, feathers, or skin) of birds and mammals tends to blend in with their background—being relative dark in warm and moist environments with their shaded undergrowth and dark soils, and relatively light in sun-exposed, rocky, sandy, or snow-covered environments. Thus, this ecogeographic pattern, exhibited by scores of species of birds and mammals across climatic regions (and, in some species from winter to summer seasons) evidences the selective advantages of crypsis (camouflage) for both predators and the prey attempting to avoid them.

The ecological and evolutionary assembly of insular biotas

The ecological drivers of natural selection and evolution are most strikingly showcased in the marvels of island life. Fundamental to understanding the forces structuring insular communities is their species-poor and **disharmonic** nature. As we learned earlier, islands are either inhabited by very few species or, if they are species-rich, their assemblages are the products of adaptive radiations from a very limited subset of lineages—those whose ancestors were among the lucky few to reach the remote, originally species-poor islands. Thus, the biota of isolated oceanic archipelagos are typically characterized as being "disharmonic" or "unbalanced"—lacking many species groups common on the mainland, but dominated by descendants of the few successful, long-distance colonists, for instance, birds and bats rather than terrestrial mammals and amphibians; flying insects rather than earthworms and other ground-dwelling invertebrates; ferns and flowering plants with wind-borne seeds rather than those more dependent on animals for dissemination of their seeds.

The marvels of island life are the products of natural selection in these highly altered evolutionary arenas, reflecting the influences of ecological release in the absence of species that typically dominate communities on the mainland (in particular, terrestrial mammals). In addition, on large and isolated islands, insular plants and animals may evolve bizarre forms in response to ecological displacement among the island's endemic species, which intensifies as the disharmonic nature of the islands increases during adaptive radiations.

The island rule—body size convergence on islands. The giant cyclops Polyphemus of Greek mythology almost certainly was inspired by a real-life, arguably equally fantastic marvel of island life. When Greeks first explored the Mediterranean islands, they discovered the bones of what appeared to be giant humans, but with large openings in the center of their skulls—the presumed eye socket of Polyphemus. Rather than some prehistoric giant, however, the skulls were actually those of dwarfs: elephants who over hundreds of generations after colonizing the islands had decreased in body size, some to less than 10% the mass of their mainland ancestors, yet still retaining the central nasal cavity that anchors the proboscis muscles of elephants.

Seemingly incredible transformations in body size characterize insular mammals both great and small. The **island rule** describes a graded trend, from dwarfism in large species of mammals and birds to gigantism in small species. This bizarre pattern is described as a "graded" trend because the degree of body size change (dwarfism or gigantism) increases from intermediate-sized species (which tend to change little in size) to those of more extreme body size. For example, the extinct insular elephant of Sicily dwarfed to just 2% of the mass of its mainland ancestor, while at the other extreme, the giant moonrat of Gargano Island (now a peninsula along the Adriatic coast of Italy) increased to over 200 times the mass of its mainland ancestor. On the mainland, competition and ecological displacement from a diverse assemblage of species drives evolutionary divergence in body size. On species-poor islands, ecological release results in reversals or relaxation in these selective pressures and convergence on intermediate body size (the island rule).

"There is no greater anomaly in nature than a bird that cannot fly." So remarked the 19th-century British paleontologist Sir Richard Owen on describing the unfathomable giant birds of New Zealand—the extinct moas, which appear to have taken on the body size and feeding habits of deer or some other large herbivore (New Zealand lacks deer and all other terrestrial mammals except for the bats). These incredible products of evolution in isolation and ecological disharmony evidence the linkage between two equally fantastic features of the island syndrome: transformations in body size (in this case gigantism in ancestrally small birds); and the loss or great attenuation in powers of dispersal (the great anomaly of "a bird that cannot fly"). The ten or so species of giant moas may have simply outgrown their wings, thus becoming giants stranded on the isolated islands of New Zealand.

The striking loss in dispersal powers of insular endemics is observed across a broad spectrum of island life forms, including hundreds of species of flightless birds, and many more flightless or nearly flightless insects. Loss or severely reduced capacities for dispersal is also common in insular plants, whose seeds have grown too large and heavy, or have lost the wing-like structures and other special adaptations for being dispersed by the winds or in the fur or feathers of mammals and birds. Once the ancestors colonized the islands, selective pressures on their descendants shifted dramatically from those for winning the ecological sweepstakes of colonizing isolated lands rich in resources and void of competing species, to minimizing the wasted loss of seeds to the surrounding open waters. As Charles Darwin put it: "As with mariners shipwrecked near a coast, it would have been better for the good swimmers if they had been able to swim still further, whereas it would have been better for the bad swimmers if they had not been able to swim at all and had stuck to the wreck." In this one metaphor, Darwin captures the reversals in natural selection that have shaped island life and, in so doing, explains Richard Owen's anomaly of birds (and insects, etc.)—descendants of long-distance dispersers, that lost the power of flight and "stuck to the wreck."

Associated with dynamics in dispersal abilities is another evolutionary marvel of island life. In comparison to trees and other woody plants, herbaceous species tend to disperse their seeds over greater distances. As a result, island colonization is typically biased in favor of herbaceous plants. But once established on islands, descendants of herbaceous plants often develop "secondary woodiness," allowing them to take on the stature of shrubs and trees. Examples include the *Opuntia* cactus trees of the Galápagos Islands, and the so-called daisy trees (an aster) of the Canary and Hawaiian Islands.

Insular plants also exhibit surprising shifts in their life histories and ecological associations. In comparison to their mainland ancestors, the timing of their flowering and seed set may be greatly altered. This may again be related to the disharmonic nature of insular communities. Isolated oceanic islands often lack the birds, bees, and mammals that typically serve as their pollinators and seed dispersers on the mainland. The plants that survive here do so by developing ecological associations with novel species such as lizards and bats which, because they may be some of the few mutualists on the islands, expand their niches and become "super pollinators" or disperse the seeds of a great variety of plant species.

On observing the island life of the Galápagos, Charles Darwin remarked that "A gun is here superfluous; for with the muzzle I pushed a hawk off the branch of a tree." Darwin, with his keen powers of observation and reasoning, was apparently the first to describe an integral and foreboding feature of the island syndrome: the tameness and ecological naivety of island endemics.

These evolutionary marvels occupy the bizarre, terminal branches of insular lineages, being shaped by natural selection in splendid isolation from mammalian carnivores, herbivores, and predators. Thus, the fauna and flora of isolated oceanic islands are often ecologically naive and highly susceptible to extinctions after mammals or other exotic and novel predators and competitors are introduced onto the islands. This explains why a highly disproportionate number of extinctions during the past 500 years have been of insular forms, and why the most astounding marvels of island life—the giant moonrats and Lilliputian hippos and elephants, and the many hundreds of species of flightless birds— all are now only found in fossil and subfossil deposits across the world's oceanic islands. This remarkable menagerie of insular beasts, many of them more bizarre than those of Greek mythology, suffered extinctions once the splendid isolation of their island homeland was lost; in most cases following colonization of even the most remote oceanic islands by our own species.

In summary, the **island syndrome** describes a suite of distinctive features of insular biotas that are exhibited across a range of scales. At the broadest scale is the disharmonic nature of insular communities—in particular, the lack or paucity of non-flying mammals that typically dominate species assemblages on the mainland. The disharmonic nature of insular communities, in turn, drives ecological dynamics at the levels of species and populations, including expansions of species' niches to occupy a broad range of habitats at super-normal population densities. Finally, at the levels of individuals, the novel selection regimes on isolated islands produce the evolutionary marvels but also the perils of island life.

Chapter 7
The geographic and ecological advance of humanity

Introspection may be one of the greatest challenges of scientific inquiry. We find it especially difficult to remain objective when analyzing the patterns and driving forces behind the actions of our own species. Yet scientists, and biogeographers in particular, should approach objectively questions relating to how our populations may have been shaped by the same selective forces that influenced other populations of wildlife as their populations expanded across the globe. And it turns out that, as much as we would like to hold ourselves above other life forms—during our rise to become the world's dominant ecological engineers—our paths of geographic expansion and subsequent patterns in geographic variation among our regional populations often parallel those of other, "less advanced" forms of native wildlife.

Global colonization of humanity

The geographic expansion of our own species out of Africa began around 90,000 years ago, and our journey of dispersal was strongly influenced by the same topographical, ecological, and climatic factors that channeled the dispersal of other species of terrestrial mammals. As they colonized new regions, the tracks of our ancestral migrations followed the warm and moist lowland regions of the tropics and subtropical latitudes—expanding eastward first across the coastal environments once early humans crossed the straits between Africa, the Arabian Peninsula, and India, most likely during glacial periods when sea levels and the extent of the straits were substantially reduced. From there, our early populations pushed rapidly eastward along the southern coasts of the Indian subcontinent, reaching Southeast Asia by about 80,000 years ago.

Colonization of the Greater Sunda Islands of Indonesia (Java, Sumatra, and Borneo) was not by swimming, rafting, or boat. This leg in our expansion into Indonesia occurred during another glacial period when sea levels dropped by some 100 meters, transforming the shallow seas off the coast of Southeast Asia into a vast land bridge joining the Greater Sunda Islands with the Asian mainland. The expansion of our geographic range eastward across the complex of Indonesian archipelagos did require some island hopping, but the gaps traversed were relatively minor (typically less than 8 km), so that by around 60,000 years ago ancestors of Australia's aboriginal people reached the glacial (low-sea level) subcontinent of Sahul, which comprised New Guinea, Australia, and Tasmania.

Colonization of islands farther east of New Guinea and Australia was much slower because these archipelagos were isolated by great expanses of open ocean. Accordingly, colonization and expansions across these regions of the Pacific Ocean required many generations of cultural evolution and the development of seaworthy vessels and advanced navigation skills. Early Pacific Islanders established populations in the Melanesian and Micronesian archipelagos by around 30,000 years ago, but didn't reach Hawaii and the other isolated archipelagoes of the central Pacific until around 2,000 years ago.

The influence of prevailing winds and ocean currents on expansions of humanity is clearly indicated by the surprisingly long delay in colonization of two large island systems, both lying relatively close to the mainland. Prevailing currents from Australia are such that they would have carried the human diaspora far north of New Zealand. The ancestors of the Maori were, thus, not descendants of native Australians but of Polynesians carried from the tropical waters of central and eastern Polynesia by the westerly flowing trade winds and ocean currents to the shores of New Zealand, which they reached just 750 years ago.

Colonization of Madagascar by the Malagasy ancestors followed an equally surprising and long-delayed path of migration. Although our populations evolved first in eastern Africa some 200,000 years ago, we didn't reach Madagascar (lying roughly 400 km off the coast of Africa) until 2,000–4,000 years ago. Any potential dispersal of early humans from Africa onto Madagascar would have been thwarted by the south-flowing Mozambique Current; so the first human colonists were not from Africa but from sources far to the east. Remarkably, modern genetic, linguistic, and archeological analyses indicate that the first Malagasy may have come from Borneo.

For the first 30,000–40,000 years of our migrations out of Africa, our ancestral populations were limited to tropical and subtropical regions of the Old World. Our eventual expansions into the colder, higher latitude and higher elevation zones of Europe and Asia awaited the cultural advances required for group hunting and foraging, engineering of effective shelters and clothing against the cold, and the use of fire. Fire is a uniquely human and unparalleled tool for stripping fur and feathers from the flesh of game animals, for providing warmth against the cold, and for managing habitats by burning away inedible, noxious, or otherwise undesirable plant life and ground cover. The Pyrenees, Alps, and Caucasus Mountains blocked our initial migrations northward into Europe, and the Himalayas blocked our migrations into the central and northern reaches of Asia; all being formidable topographic and climatic barriers. Eventually, our early civilizations would conquer the cold, cross these barriers, and reach the western shores of Europe by around 45,000 years ago, and the northern regions of Asia and into present-day Siberia by around 25,000 years ago.

Again, glacial cycles of the Pleistocene were fundamental to our continued expansion—this time into the New World. Our expansion into North America was via a two-stage colonization. The first stage occurred during a glacial period when sea levels dropped by about 100 meters, transforming the shallow seas between Siberia and Alaska into the glacial subcontinent of Beringia, which persisted between 36,000 and 16,000 years ago. Because of the wind patterns that prevailed during that period, Beringia remained ice-free and harbored a great abundance of plants and a diversity of megafauna and other wildlife that may have rivaled that of today's African savannahs.

The second stage of human expansion into the New World would await the initial melting of vast glaciers that stretched across the extent of northern North America and blocked our southward expansion. By around 13,000–15,000 years ago, passages opened along the Pacific coast and along a central, inland corridor between two retreating ice sheets to the west and east. Once gaining footholds in North America south of the remnant and waning glaciers, our expansion across the rich, open expanses of the continent were remarkably rapid—reaching both coasts and then southward into present-day Mexico and down into Central America by around 12,000 years ago. While the Andes of eastern South America likely impeded migrations east to west, their N–S orientation allowed rapid expansions southward to the extreme southern tip of South America—an astonishing surge of advancing, aboriginal Americans across the 15,000 km extent of the western hemisphere from Beringia to Tierra del Fuego in just 2,000 years.

Natural selection and ecogeography of indigenous humans

As our ancestral populations spread across the globe, they encountered the same environments and selective forces that shaped the phenotypes of other populations endemic to these regions. The imprint of these regional selective forces is still evidenced by genetically determined differences in the characteristics of indigenous populations of humans, who often exhibit the same ecogeographic patterns described earlier for other native wildlife.

In comparison to populations endemic to tropical regions of the continents, human populations native to lands in the higher latitudes tend to be larger and have relatively shorter limbs—following Bergmann's and Allen's rules, respectively. Our indigenous populations also exhibit latitudinal variation in skin color reminiscent of Gloger's rule: darker skin (greater melanism) in the tropics, protecting native humans from the potentially destructive effects of intense solar radiation; lighter skin in high latitude populations of our species promoting absorption of the limited sunlight to levels sufficient to stimulate the production and storage of vital nutrients in the skin.

On islands, the body size of primates (including insular populations of hominids) generally follows the island rule—exhibiting a graded trend toward more pronounced dwarfism in the larger species. The hominid of Flores Island (the "hobbit") was only one-third to one-half the mass of its ancestor (most likely, *Homo erectus*). The island peoples of Indonesia, Melanesia, Micronesia, and the Philippines also tend to be relatively small in stature, again consistent with the island rule.

Islanders native to the more isolated archipelagos of the Pacific, however, tend to be exceptionally large. We can resolve this apparent anomaly by recalling Darwin's insightful metaphor about shipwrecked mariners. Selection for those who can "swim still further", also termed **immigrant selection**, should yield colonizing populations that are biased in favor of the more powerful dispersers. Therefore, because larger mammals have more energy stores relative to how much energy they need to travel a given distance, the body size of insular primates (including humans) should increase with archipelago isolation. These selective pressures were likely compounded for lineages of Pacific islanders, where immigrant selection (or what anthropologists refer to as "selection for the thrifty genotypes") was a recurrent process—essential for their long-term survival, which depended on continuously migrating among distant archipelagos and exploring new ones to hedge against eventual collapse of their limited island resources.

An epilogue of extinction and homogenization

The final chapter in the story of the early, global expansion of humanity is a poignant and sobering one. Whereas wildlife of Africa coevolved with our ancestral populations, adapting to each incremental advance in our species' ecological prowess, those of other lands were naive to what Darwin warned was the "strangers' craft of power." The megafaunal and other native mammals, birds, and reptiles were woefully incapable of adapting to the accumulated powers of these novel hunters and ecosystem engineers, falling like dominoes on the heels of our arrival in these "new" lands. Prehistoric waves of extinctions occurred first in Australia, then Europe and Asia, followed by North America, South America, and then on to islands of the Mediterranean and Caribbean, following in lock-step the same sequence of human colonizations across the globe.

The more recent, historic waves of extinctions of megafauna and other ecologically naive wildlife on oceanic islands followed the tract of colonizations by Pacific Islanders—ultimately exterminating at least seventeen of Madagascar's largest lemurs and all of the ten or so species of New Zealand's giant flightless birds—the moas. The saga of anthropogenic extinctions was repeated countless times across the marine realm as hundreds, and possibly thousands of species of flightless birds and other insular endemics suffered extinctions at the hands, teeth, and claws of our colonizing populations and the legion of rats, mice, cats, weasels, goats, and other species we introduced to even the most remote islands. Not only did we severely reduce the distinctiveness of oceanic islands by driving thousands of endemic species to extinction, but we compounded this by introducing a redundant suite of commensal species to these islands. The result was a global-scale homogenization of nature; a dissolution of biogeography's most fundamental pattern, Buffon's Law—the biological distinctiveness of place.

The promise of a synergy: conservation biogeography

Rather than conclude in pathos lamenting over anthropogenic extinctions, it seems imperative that we marshal our innate connection with what remains of the marvelous menagerie of native life forms and apply the principles and tools of biogeography to conserve them. E. O. Wilson's observation that biogeography had become the cornerstone of conservation biology is reaffirmed and now codified in the relatively recent articulation of a synergy of these two disciplines—the emerging field of **conservation biogeography**. One of the principal tenets of this synergy of the theoretical and the applied is that successful strategies for conserving the diversity and natural character of native biotas require that:

(1) we address what is known as the **Wallacean shortfall**—the vast gap in our knowledge of the distributions of species and the geographic dynamics of the forces threatening their survival; and

(2) we conserve the distinctive character of native species, by conserving the geographic context of their native populations. As valuable as zoos and nature reserves have been for maintaining individuals of particular species, they fall woefully short of recreating the ecological and selective regimes that molded animal and plant lineages over their evolutionary histories.

Elephants persisting in captivity and across the shrinking and highly fragmented remnants of their native range provide a poignant case study in the potential waning of the natural character of wildlife now persisting under highly altered selective regimes. African and Asian elephants have been subjected to centuries of anthropogenic isolation, hunting, and slaughter by the ivory trade (in both cases, selecting out the largest individuals), and conscription to serve as beasts of burden or as performers in circuses and exhibits in zoos (each of these often weeding out the most unruly or "wildest" individuals). The combined effects of these anthropogenic but persistent selection pressures may be the transformation of the Earth's largest and once most dominant species of terrestrial wildlife into downsized and relatively tame species, with some populations of "wild" elephants becoming genetically fixed on tusklessness, thus losing another one of their most defining features.

It seems especially important that I conclude this book on the insights and inspirations from the geography of life with a note of scientific humility and of optimism. Omniscience is not the goal of science, and those who pretend to have achieved such are destined to stagnation. The most promising approach for advancing our understanding of the natural world is more likely to be to admit and embrace scientific ignorance—rigorously and objectively searching for the gaps in our individual and collective knowledge. The so-called "maps of ignorance" being developed by Joaquin Hortal and his colleagues (Figure 32) are exemplary creations of a new and critical line of research that can strategically focus scientific explorations to address the Wallacean shortfall and

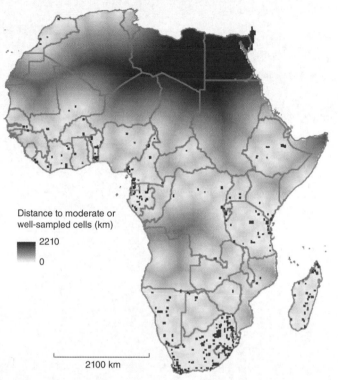

Distance to moderate or
well-sampled cells (km)

2210

0

2100 km

32. Maps of ignorance can provide especially important insights on
the Wallacean shortfall (i.e. gaps in our knowledge of the geography
of life), enabling us to strategically target our limited time, energy,
and resources to those regions and biotas in most need of
biogeographic study.

provide the data essential for conserving the diversity and natural
character of native biotas. The frontiers of biogeography in
general, and of conservation biogeography in particular, are being
defined by scientists who are developing novel approaches for
identifying and visualizing the critical gaps in our knowledge of
the geography of life.

137

References and further reading

Chapter 1: Biological diversity and the geography of nature

Darwin, C. 1859. *On the Origin of Species by Means of Natural Selection or the Preservation of Favored Races in the Struggle for Life*. John Murray.

Ellis, E. C. 2018. *Anthropocene: A Very Short Introduction*. Oxford University Press.

Humboldt, A. von and A. Bonpland. 2009. *Essay on the Geography of Plants* (1807). Trans. S. Romanowski; ed. S. T. Jackson; accompanying essays and supplementary material by S. T. Jackson and S. Romanowski. University of Chicago Press.

Jackson, S. T. and L. D. Walls. 2014. *Views of Nature: Alexander von Humboldt*. University of Chicago Press.

Lomolino, M. V., B. R. Riddle, and R. J. Whittaker. 2017. *Biogeography*, 5th Edition. Sinauer Press.

Lomolino, M. V., D. F. Sax, and J. H. Brown (eds) 2004. *Foundations of Biogeography*. University of Chicago Press.

McCarthy, D. 2009. *Here Be Dragons: How the Study of Animal and Plant Distributions Revolutionized our Views of Life and Earth*. Oxford University Press.

Wallace, A. R. 1876. *The Geographical Distribution of Animals*. 2 vols. Macmillan.

Wilson, E. O. 1994. *Naturalist*. Island Press.

Winchester, S. 2001. *The Map that Changed the World: William Smith and the Birth of Modern Geology*. Harper Collins.

Wulf, A. 2015. *The Invention of Nature: Alexander von Humboldt's New World*. Alfred A. Knopf.

Chapter 2: Dynamic maps of a dynamic planet

Hess, H. H. 1962. History of ocean basins. In A. E. J. Engel, H. L. James, and B. F. Leonard (eds), *Petrological Studies: A Volume in Honor of A. F. Buddington*, 599–620. Geological Society of America.

Martin, P. S. 1967. Prehistoric overkill. In P. S. Martin and H. E. Wright Jr (eds), *Pleistocene Extinctions: The Search for a Cause*, 75–120. Yale University Press.

Merriam, C. H. 1892. The geographical distribution of life in North America with special reference to the Mammalia. *Proceedings of the Biological Society of Washington* 7: 1–64.

Pielou, E. C. 1991. *After the Ice Age*. University of Chicago Press.

Sandom, C., S. Faurby, B. Sandel, and J.-C. Svenning. 2014. Global late Quaternary megafauna extinctions linked to humans, not climate change. *Proceedings of the Royal Society* B 281 20133254. DOI: 10.1098/rspb.2013.3254.

Snider-Pellegrini, A. 1858. *La Création et ses mysteres*. Frank and Dentu.

Wegener, A. 1966. *The Origin of Continents and Oceans*. Dover Publications. [Translation of 1929 edition by J.Biram.]

Chapter 3: The geography of diversification

Brawand, D., et al. 2014. The genomic substrate for adaptive radiation in African cichlid fish. *Nature* 513: 375–81. doi:10.1038/nature13726.

Fryer, G. and T. D. Iles. 1972. *The Cichlid Fishes of the Great Lakes of Africa: Their Biology and Evolution*. Oliver & Boyd.

Gillespie, R. G. 2015. Island time and the interplay between ecology and evolution in species diversification. *Evolutionary Applications* ISSN 1752-4571—doi:10.1111/eva.12302.

Givnish, T. J., K. C. Millam, A. R. Mast, T. B. Paterson, T. K. Themi, A. L. Hipp, J. M. Henss, J. F. Smith, K. R. Wood, and K. J. Sytsma. 2009. Origin, adaptive radiation and diversification of the Hawaiian lobeliads (Asterales: Campanulaceae). *Proceedings of the Royal Society* B: *Biological Sciences* 276: 407–16.

Goodman, S. M. and J. P. Benstead. 2004. *The Natural History of Madagascar*. University of Chicago Press.

Grant, P. R. 1986. *Ecology and Evolution of Darwin's Finches*. Princeton University Press.

Losos, J. B. and R. E. Ricklefs. 2009. Adaptation and diversification on islands. *Nature* 457: 830–6.

Mittermeier, R. A., F. Hawkins, and E. E. Louis. 2010. *Lemurs of Madagascar*, 3rd Edition. Arlington, VA: Conservation International.

Wagner, W. L. and V. A. Funk (eds). 1995. *Hawaiian Biogeography: Evolution on a Hot Spot Archipelago*. Smithsonian Institution Press.

Sources for Table 1

Ali, J. R. and M. Huber. Mammalian biodiversity on Madagascar controlled by ocean currents. *Nature*. Nature Publishing Group. 463 (Feb. 4, 2010): 653–6. Bibcode:2010Natur.463..653A. doi:10.1038/nature08706. PMID 20090678. Retrieved Jan. 20, 2010.

Buerki, S., D. S. Devey, M. W. Callmander, P. B. Phillipson, and F. Forest. 2013. Spatio-temporal history of the endemic genera of Madagascar, *Botanical Journal of the Linnean Society* 171 (2) (February): 304–29, https://doi.org/10.1111/boj.12008.

Callmander, M. et al. (2011). The endemic and non-endemic vascular flora of Madagascar updated. *Plant Ecology and Evolution*. 144 (2): 121–5. doi:10.5091/plecevo.2011.513.

Gehring, P.-S., J. Kohler, A. Straub, R. D. Randrianiaina, J. Glos, F. Glaw, and M. Vences. 2011. The kingdom of the frogs: anuran radiations in Madagascar. In F. E. Zachos and J. C. Habel (eds), *Biodiversity Hotspots*, DOI 10.1007/978-3-642-20,992-5_13, # Springer-Verlag.

Goodman, S. M. and J. P. Benstead. 2005. Updated estimates of biotic diversity and endemism for Madagascar. *Oryx* 39: 73–7.

Kinver, M. Mammals "floated to Madagascar". BBC News website. BBC. Retrieved Jan. 20, 2010.

Mittermeier, R. A., P. R. Gil, M. Hoffman, J. Pilgrim, T. Brooks, C. G. Mittermeier, J. Lamoreux, and G. A. B. da Fonseca 2004. *Hotspots Revisited*. Chicago University Press.

Mittermeier, R. A., F. Hawkins, and E. E. Louis. 2010. *Lemurs of Madagascar*, 3rd Edition. Conservation International.

Mittermeier, R., J. Ganzhorn, W. Konstant, K. Glander, I. Tattersall, C. Groves, A. Rylands, A. Hapke, J. Ratsimbazafy, M. Mayor, E. Louis, Y. Rumpler, C. Schwitzer, and R. Rasoloarison. 2008. Lemur diversity in Madagascar. *International Journal of Primatology* 29 (6): 1607–56.

Nagy, Z. T., U. Joger, M. Wink, F. Glaw, and M. Vences. 2003. Multiple colonization of Madagascar and Socotra by colubrid snakes: evidence from nuclear and mitochondrial gene phylogenies. *Proceedings. Biological Sciences* 270 (1533): 2613–21.

Rakotondrainibe, F. 2003. Checklist of the pteridophytes of Madagascar. In S. M. Goodman and J. P. Benstead (eds), *Natural History of Madagascar*, 295–313. Chicago University Press.

Raselimanana, A. P., B. Noonan, K. K. Praveen, J. Gauthier, and A. Yoder. 2008. Phylogeny and evolution of Malagasy plated lizards. *Molecular Phylogenetics and Evolution* 50: 336–44. 10.1016/j.ympev.2008.10.004.

Raxworthy, C. J. 2003. Introduction to the reptiles. In S. M. Goodman and J. P. Benstead. *The Natural History of Madagascar*, 934–49. University of Chicago Press.

Samonds, K. E. 2012. Spatial and temporal arrival patterns of Madagascar's vertebrate fauna explained by distance, ocean currents, and ancestor type. *PNAS* 109 (April 3): 5352–7.

Stiassny, M. L. J. 1992. Phylogenetic analysis and the role of systematics in the biodiversity crisis. In N. Eldredge (ed.), *Systematics, Ecology and the Biodiversity Crisis*, 109–20. Columbia University Press.

The Reptile Database; Reptiles of Madagascar <http://reptile-database. reptarium.cz/advanced_search?location=Madagascar&submit =Search>

Vieites, D. R., K. C. Wollenberg, F. Andreone, J. Köhler, F. Glaw, and M. Vences. 2009. Vast underestimation of Madagascar's biodiversity evidenced by an integrative amphibian inventory. *Proceedings of the National Academy of Sciences* 106 (20): 8267–72.

Chapter 4: Retracing evolution across space and time

Avise, J. C. 2000. *Phylogeography: The History and Formation of Species*. Harvard University Press.

Browne, J. 1983. *The Secular Ark: Studies in the History of Biogeography*. Yale University Press.

Cowie, R. H., K. A. Hayes, C. T. Tran, and W. M. Meyer, III. 2008. The horticultural industry as a vector of alien snails and slugs: widespread invasions in Hawaii. *International Journal of Pest Management* 54: 267–76.

Haeckel, E. 1876 (later editions in 1907, 1911). *The History of Creation, or, The Development of the Earth and its Inhabitants by the Action of Natural Causes: A Popular Exposition of the Doctrine of Evolution in General and of that of Darwin, Goethe and Lamarck in Particular*. D. Appleton.

Kreft, H. and W. Jetz. 2010. A framework for delineating biogeographical regions based on species distributions. *Journal of Biogeography* 37: 2029–53.

McKinney, M. and J. Lockwood (eds). 2001. *Biotic Homogenization: The Loss of Diversity through Invasion and Extinction*. Kluwer Academic/Plenum Publishers.

Poulakakis, N., M. Russello, D. Geist, and A. Caccone. 2012. Unravelling the peculiarities of island life: vicariance, dispersal and the diversification of the extinct and extant giant Galápagos tortoises. *Molecular Ecology* 21: 160–73.

Sax, D. F., J. J. Stachowicz, and S. D. Gaines. 2005. *Species Invasions: Insights into Ecology, Evolution and Biogeography*. Sinauer Associates.

Shapiro, L. H., J. S. Strazanac, and G. K. Roderick. 2006. Molecular phylogeny of Banza (Orthoptera: Tettigoniidae), the endemic katydids of the Hawaiian Archipelago. *Molecular Phylogenetics and Evolution* 41: 53–63.

Takhtajan, A. 1986. *Floristic Regions of the World*. University of California Press.

Wallace, A. R. 1876. *The Geographical Distribution of Animals*. 2 volumes. Macmillan.

Chapter 5: The geography of biological diversity

Diamond, J. M. 1975. Assembly of species communities. In M. L. Cody and J. M. Diamond (eds), *Ecology and Evolution of Communities*, 342–444. Belknap Press.

Forster, J. R. 1778. *Observations Made during a Voyage Round the World, on Physical Geography, Natural History and Ethic Philosophy*. G. Robinson.

Heaney, L. R. and J. Regalado. 1998. *Vanishing Treasures of the Philipipine Rainforests*. University of Chicago Press.

Lomolino, M. V. 2000. Ecology's most general, yet protean pattern: the species–area relationship. *Journal of Biogeography* 27: 17–26.

MacArthur, R. H. and E. O. Wilson. 1967. *The Theory of Island Biogeography*. Princeton University Press.

Mutke, J. and W. Barthlott. 2005. Patterns of vascular plant diversity at continental to global scales. *Biologiske Skrifter* 55: 521–31.

Schipper, J. and 129 others. 2008. The status of the world's land and marine mammals: diversity, threat and knowledge. *Science* 322: 225–30. doi: 10.1126/science.1165115.

Simberloff, D. S. and E. O. Wilson. 1969. Experimental zoogeography of islands: the colonization of empty islands. *Ecology* 50: 278–96.

Simberloff, D. S. and E. O. Wilson. 1970. Experimental zoogeography of islands: a two-year record of colonization. *Ecology* 51: 934–7.

Whittaker, R. J. and J. M. Fernández-Palacios. 2007. *Island Biogeography: Ecology, Evolution, and Conservation*, 2nd Edition. Oxford University Press.

Whittaker, R. J., K. A. Triantis, and R. J. Ladle. 2008. A general dynamic theory of oceanic island biogeography. *Journal of Biogeography* 35: 977–94.

Wilson, E. O. 1986. *Biophilia*. Harvard University Press.

Chapter 6: Macroecology and the geography of micro-evolution

Allen, J. A. 1878. The influence of physical conditions in the genesis of species. *Radical Review* 1: 108–40.

Bergmann, C. 1847. Über die Verhältnisse der Wärmeökonomie der Thiere zu ihren Grösse. *Göttinger Studien* 1: 595–708.

Brown, J. H. 1995. *Macroecology*. University of Chicago Press.

Carlquist, S. 1974. *Island Biology*. Columbia University Press.

Gloger, C. L. 1883. *Das Abandern der Vogel durch Einfluss des Klimas*. A. Schulz.

Jordan, D. S. 1891. *Temperature and Vertebrae: A Study in Evolution*. Wilder-Quarter Century Books.

Rapoport, E. H. 1982. *Areography: Geographical Strategies of Species*. Pergamon Press.

Chapter 7: The geographic and ecological advance of humanity

Bellwood, P. (ed.) 2013. *The Global Prehistory of Human Migration*. Wiley-Blackwell.

Finlayson, C. 2005. Biogeography and evolution of the genus Homo. *Trends in Ecology and Evolution* 20: 457–63.

Hortal, J., D. Rocchini, S. Lengyel, J. M. Lobo, A. Jiménez-Valverde, C. Ricotta, G. Bacaro, and A. Chiarucci. 2011. Accounting for uncertainty when mapping species distributions: the need for maps of ignorance. *Progress in Physical Geography* 35: 221–6.

Howells, W. 1973. *The Pacific Islanders*. Weidenfeld and Nicolson.

Oppenheimer, C. 2003. *Out of Eden: The Peopling of the World*. Constable.

Terrell, J. 1986. *Prehistory in the Pacific Islands: A Study of Variation in Language, Customs and Human Biology*. Cambridge University Press.

Wilson, E. O. 1999. Prologue. In G. Daws and M. Fujita (eds), *Archipelago: The Islands of Indonesia*. University of California Press.

Index

Index

CANCER
A Very Short Introduction
Nick James

Cancer research is a major economic activity. There are constant improvements in treatment techniques that result in better cure rates and increased quality and quantity of life for those with the disease, yet stories of breakthroughs in a cure for cancer are often in the media. In this *Very Short Introduction* Nick James, founder of the CancerHelp UK website, examines the trends in diagnosis and treatment of the disease, as well as its economic consequences. Asking what cancer is and what causes it, he considers issues surrounding expensive drug development, what can be done to reduce the risk of developing cancer, and the use of complementary and alternative therapies.

THE HISTORY OF MEDICINE

A Very Short Introduction

William Bynum

Against the backdrop of unprecedented concern for the future of health care, this Very Short Introduction surveys the history of medicine from classical times to the present. Focussing on the key turning points in the history of Western medicine, such as the advent of hospitals and the rise of experimental medicine, Bill Bynum offers insights into medicine's past, while at the same time engaging with contemporary issues, discoveries, and controversies.

www.oup.com/vsi

Superconductivity

A Very Short Introduction

Stephen J. Blundell

Superconductivity is one of the most exciting areas of research in physics today. Outlining the history of its discovery, and the race to understand its many mysterious and counter-intuitive phenomena, this *Very Short Introduction* explains in accessible terms the theories that have been developed, and how they have influenced other areas of science, including the Higgs boson of particle physics and ideas about the early Universe. It is an engaging and informative account of a fascinating scientific detective story, and an intelligible insight into some deep and beautiful ideas of physics.

THE HISTORY OF LIFE
A Very Short Introduction
Michael J. Benton

There are few stories more remarkable than the evolution of life on earth. This *Very Short Introduction* presents a succinct guide to the key episodes in that story - from the very origins of life four million years ago to the extraordinary diversity of species around the globe today. Beginning with an explanation of the controversies surrounding the birth of life itself, each following chapter tells of a major breakthrough that made new forms of life possible: including sex and multicellularity, hard skeletons, and the move to land. Along the way, we witness the greatest mass extinction, the first forests, the rise of modern ecosystems, and, most recently, conscious humans.

SLEEP
A Very Short Introduction
Russell G. Foster & Steven W. Lockley

Why do we need sleep? What happens when we don't get enough? From the biology and psychology of sleep and the history of sleep in science, art, and literature; to the impact of a 24/7 society and the role of society in causing sleep disruption, this *Very Short Introduction* addresses the biological and psychological aspects of sleep, providing a basic understanding of what sleep is and how it is measured, looking at sleep through the human lifespan and the causes and consequences of major sleep disorders. Russell G. Foster and Steven W. Lockley go on to consider the impact of modern society, examining the relationship between sleep and work hours, and the impact of our modern lifestyle.

SEXUALITY
A Very Short Introduction
Veronique Mottier

What shapes our sexuality? Is it a product of our genes, or of society, culture, and politics? How have concepts of sexuality and sexual norms changed over time? How have feminist theories, religion, and HIV/AIDS affected our attitudes to sex? Focusing on the social, political, and psychological aspects of sexuality, this *Very Short Introduction* examines these questions and many more, exploring what shapes our sexuality, and how our attitudes to sex have in turn shaped the wider world. Revealing how our assumptions about what is 'normal' in sexuality have, in reality, varied widely across time and place, this book tackles the major topics and controversies that still confront us when issues of sex and sexuality are discussed: from sex education, HIV/AIDS, and eugenics, to religious doctrine, gay rights, and feminism.

Memory
A Very Short Introduction
Michael J. Benton

Why do we remember events from our childhood as if they happened yesterday, but not what we did last week? Why does our memory seem to work well sometimes and not others? What happens when it goes wrong? Can memory be improved or manipulated, by psychological techniques or even 'brain implants'? How does memory grow and change as we age? And what of so-called 'recovered' memories? This book brings together the latest research in neuroscience and psychology, and weaves in case-studies, anecdotes, and even literature and philosophy, to address these and many other important questions about the science of memory - how it works, and why we can't live without it.